6

TRAFFIC FLOW
ON TRANSPORTATION
NETWORKS

MIT Press Series in Transportation Studies

Marvin L. Manheim, editor
Center for Transportation Studies, MIT

TRAFFIC FLOW
ON TRANSPORTATION
NETWORKS

G. F. Newell

The MIT Press
Cambridge, Massachusetts,
and London, England

This book was set in Monophoto Times Mathematics by Asco Trade Typesetting Limited, Hong Kong, printed and bound by Halliday Lithograph Corporation in the United States of America

Library of Congress Cataloging in Publication Data

Newell, Gordon Frank, 1925–
    Traffic flow on transportation networks.

    (MIT Press series in transportation studies; 5)
    Includes bibliographies and index.
    1. Traffic flow—Mathematical models. 2. Transportation planning—Mathematical models. I. Title. II. Series.
HE336.T7N48   380.5′07′24   79-28843
ISBN 0-262-14032-2

CONTENTS

SERIES FOREWORD

The field of transportation has emerged as a recognized profession only in the last ten years, although transportation issues have been important throughout history. Today, more and more government agencies, universities, researchers, consultants, and private industry groups are becoming truly multi-modal in their orientations, and specialists of many different disciplines and professions are working on multi-disciplinary approaches to complex transportation issues.

The central role of transportation in our world today and its recent professional status have led The MIT Press and the MIT Center for Transportation Studies to establish The MIT Press Series in Transportation Studies. The series will present works representing the broad spectrum of transportation concerns. Some volumes will report significant new research, while others will give analyses of specific policy, planning, or management issues. Still others will show the interaction between research and policy. Contributions will be drawn from the worldwide transportation community.

This book, the fifth in the series, deals with the quantitative analysis of transportation systems. The author, Gordon F. Newell, focuses on the important issue of the prediction of flows in transportation networks.

Marvin L. Manheim

PREFACE

This book is, essentially, the lecture notes for a course given to transportation engineering students at the University of California, Berkeley. A course with this title was first given in 1967 jointly by Renfrey Potts and Robert Oliver when Potts was a visiting professor on leave from the University of Adelaide. Potts and Oliver had for some years been reviewing the literature on transportation planning models and were appalled by some of the methods used and the high proportion of mathematically incorrect procedures. One of their main purposes was to present a mathematically correct and coherent account of the existing procedures.

The notes from this original course were later used separately by Potts and Oliver in other courses. Potts returned to Adelaide and gave lectures to applied mathematics students. Oliver returned to his own department to use the notes in some operations research courses. The notes later evolved into the often-quoted book *Flows in Transportation Networks* (Academic Press, 1972).

The year after Potts and Oliver gave the course and went their separate ways, the burden of continuing the course fell upon me. At the time I knew little about the subject beyond what I had learned from attending the lectures the previous year and from various seminars. I was not even particularly interested in the subject. Although I could appreciate the efforts Potts and Oliver had made to sort out the literature and make a coherent account of it, I had a feeling that the result represented some elegant answers to the wrong questions.

In 1968 the "transportation planning procedures" were still very popular; a number of major studies were in progress, including one of the San Francisco Bay Area. My first attempt at giving this course was partly a reinterpretation of the Potts-Oliver notes, partly a critique of the modeling procedures, and partly some reinterpretations of the studies of idealized networks done in England by Reuben Smeed and collaborators. My first set of notes was in outline not much different from the present set although the notes have been rewritten and expanded several times. They are now so different from the Potts-Oliver book, however, that one would hardly recognize that they are notes from the same course (the catalogue description of which has hardly changed).

Many of the views expressed here were rather unpopular in 1968, when several transportation studies were still in progress. Although people still are making such studies, the critics are now more numerous than the supporters. The vogue has now shifted to "demand models" (which are not obviously any more logical).

This book is not really intended to be a critique of the transportation planning procedures; it is even narrower in scope than the Potts-Oliver book. There is no discussion of "distribution models," which occupies a sizable part of the Potts-Oliver book. It is left out because I believe that whoever predicts future land use and, from that, the generation of trips (by guesswork and instinct), can probably also guess where the trips will go at least as well as any gravity (logit, or whatever) model. The text is essentially limited to the "assignment problem," and how the flow pattern generated from a given trip table depends on the properties of a network.

I have delayed publishing these notes for several years because I have felt that they also may give coherent answers to the wrong questions (although I do believe, after ten years, that I may be a little closer to the right questions.) Since, however, it is clear that I will not soon produce answers to most of the questions that I and others have raised, it seems time to publish a progress report and hope that it will stimulate others to make it obsolete.

If the reader is frustrated after reading this book because he does not understand the "correct" way to plan a transportation network, he will only share my feelings. Perhaps he will also share my skepticism about conventional approaches.

I am indebted to Renfrey Potts and Robert Oliver for supplying the seeds from which the book grew. I also benefited from many discussions with Gerald Steuart in planning the original outline. The greatest stimulus, however, comes from the many students whom I have tormented with this.

The typing of multiple versions of these notes was done by Phyllis De Fabio. Most of the final version was typed by Inta Vodopals.

OUTLINE

Chapter 1 gives a very brief outline of the conventional transportation planning procedures. Although subsequent chapters deal with only narrow aspects of the whole problem, it is important first to understand how the results of the theory are to be applied. The main thesis is that traffic assignment schemes with very poor real or predicted data will be used for making crude estimates of traffic flow on various proposed networks. It is, therefore, more important to have a theory that helps one to understand basic issues than to have a theory that gives accurate results; that is, if the theory and the data were accurate. The ultimate problem is to interpret the consequences of making changes in an existing network to accommodate future demands.

The modeling of a network is presented in several stages. Chapter 2 deals with elementary definitions and properties of graphs, particularly with the notion of routes and accessibility. At this stage, the graph has no properties other than those associated with the existence or nonexistence of links between various nodes. In chapter 3 a metric on nodes of the graph is introduced as a representation of geometric distance or travel time. This is a function defined on pairs of nodes. Particular emphasis, however, is placed on the metrics generated from shortest paths on a graph for which there is a distance defined on links.

Chapter 4 describes, in some detail, the metric (travel time) on networks of various idealized geometries (rectangular grids, ring-radial networks, etc.), but, particularly, on networks consisting of a single fast route superimposed on a fine network of slow-speed routes. The thesis here is that in some future generation of traffic assignment models one should exploit certain obvious geometric properties of real transportation networks rather than use algorithms that apply to any abstract graph. One important property of real transportation networks is that they are nearly planar graphs embedded in an Euclidean space. They also have a hierarchical structure consisting of coarse networks of fast routes superimposed on finer networks of slower routes. Although the flows on the minor roads are not of particular interest in themselves (they are not usually congested), one does wish to have a realistic estimate of how much "background" flow is generated on the fast routes by short trips that use the fast roads for only a small part of their trip. If one could estimate these flows by analyzing local traffic patterns, one could then investigate the routing of the long trips.

Chapter 5 deals with the definition of flows on networks and various mathematical representations of the conservation equations. The main emphasis is on the relation between link flows and route flows and the representation of flows as the superposition of various component flows. Flows on a fine grid are also described in terms of fluxes on a two-dimensional continuum.

In chapter 6, the separate notions of a metric and a flow introduced in chapters 3 and 5 are combined. We now consider the specific flow patterns generated from a given trip table if trips are assigned to cheapest paths (in the sense of the given cost metric of chapter 3) or are assigned so as to minimize the total travel cost for all trips. This chapter includes a discussion of uniqueness of the assignments and methods for evaluating the flows, particularly when the travel costs depend on the flows.

Chapter 7 deal with some of the idealized networks of chapter 4. It is first shown that if a traffic assignment procedure minimizes some convex function of the flows and the traffic assignment problem maps into itself under certain symmetry transformations, then the resulting flow pattern shows the same symmetry. This can be exploited to determine the flows on certain idealized networks, particularly rectangular grids with translational symmetry in two directions. These solutions demonstrate how flows will distribute themselves over a network in a manner that is insensitive to the details of the trip length distribution; the flows are essentially determined by the total vehicle-miles of travel in various directions.

Even if the roads in a rectangular grid are not exactly the same or the O-D table does not have complete translational symmetry, the assignment of trips to minimize individual trip times tends to give equal cost per unit distance (or speed) of travel on all parallel roads. The flows are still quite insensitive to the O-D table. In analyzing flows in a downtown region of a city where there may be a regular grid of roads for reasons such as land use or access, one should exploit this fact in determining the assignments and flows.

Chapter 8 deals with "nonidentical travelers;" it covers possible consequences of the fact that different people will interpret the costs of various routes differently. If each person chooses what he thinks is the cheapest route, he will derive a greater benefit than if he is assigned to the route that has the cheapest average cost, averaged over all travelers. It is, therefore, advantageous to make routes between the same origin and destination different in any way so that travelers will disagree on the merits of the routes. This also implies that the networks with the idealized geometry described in previous chapters are not likely to be the most efficient.

Chapter 9 contains a brief introduction to the question of what should be built to accommodate trips. If construction cost is added to the transportation cost, the economy of scale in construction converts the convex traffic assignment problem into a nonconvex problem. In essence, the goal is to balance the increased efficiency gained from focusing trips onto the same links against the increased trip length necessary to focus the trips. It is shown that an efficient network should have a hierarchical structure, and it is likely to have a nearly rectangular geometry locally.

TRAFFIC FLOW
ON TRANSPORTATION
NETWORKS

1 TRANSPORTATION PLANNING

1.1 Introduction

The goal of transportation planning is to make the best possible prediction of future transportation needs and the consequences (benefits, costs, and so forth) of the construction of various types of transportation facilities. This goal touches on many disciplines ranging from science to politics, including economics, city planning, sociology, and engineering. Even a well-educated person can know only a small part of what is relevant to the analysis of fundamental problems.

Almost all aspects of transportation planning involve numbers and logic, therefore mathematics. We will consider here only a rather narrow range of problems that will illustrate some applications of mathematics. The main emphasis will be on the translation of certain concepts into a mathematical language, and vice versa, and how one can use mathematics to illustrate and clarify some basic issues in transportation network design.

The main qualitative difference between the class of mathematics problems encountered in planning and those encountered in more conventional engineering fields is that, in planning, the accuracy of predictions is inherently limited by uncertainty in social values, population growth, technology changes, and so forth. Whereas in certain branches of engineering, a crude estimate is considered to be what a slide rule can give, in transportation planning one is often pleased if one can estimate something to within a factor of 2.

Transportation systems are usually very complicated and involve an enormous number of variables. The mathematical art of analysis is largely one of sifting out the relevant from the irrelevant and reducing what, in principle, is a hopelessly involved problem to one that is manageable or perhaps even trivial with a level of accuracy consistent with basic limitations in knowledge. This requires special mathematical or engineering skills quite different from those one learns in more conventional mathematics or engineering courses.

The "transportation planning" procedures could presumably be applied to (1) building a single road to serve the traffic that will exist during the next one to five years (a time comparable with the engineering design and construction period), (2) planning the expansion of a transportation system for a city over a period of about twenty years, or (3) designing transportation facilities for future new cities. The basic principles that one should apply to problems of various sizes should presumably be similar, but in practice they must be quite different. Problems of small size with short horizons can be analyzed quite accurately. Problems of large size with long horizons must be analyzed skillfully and crudely.

Over the last twenty or thirty years, a scheme of analysis has evolved through such studies as the Chicago Transportation Study, the Penn-Jersey Study, the Bay Area Transportation Study, and the Northeast Corridor Study. Some of these procedures have become quite stereotyped. In the United States, much of the funding comes from the federal government, and these studies have had to conform with certain specifications formulated by the government through the Department of Transportation. Nearly every large traffic consulting firm is equipped to make a study of any metropolitan region following procedures initiated by the 1962 Federal Highways Act.

The main features of this procedure are outlined briefly not because we intend to examine it critically piece by piece or in its entirety but because examination of any part of it should be done in the context of its intended use. In particular, it is the parts of the planning that we do *not* intend to examine carefully that impose inherent limitations on the overall accuracy and thereby dictate the accuracy with which any one part should be analyzed. Although rather little is known about the accuracy of past applications of the overall procedures, it is quite clear that it could not possibly be very high.

1.2 Outline of Procedure

The four-phase transportation planning procedure is essentially a procedure that evolved during the 1960s. The details of certain parts of it have changed somewhat during the

1970s, but much of the basic philosophy has survived. Although this outline may not quite correspond to current views, it still represents a fair description of the procedure and will serve at least as a background for the analysis in succeeding chapters. Since these methods were extremely expensive to implement, recent trends have often involved the use of "models" to replace the collection of real data, a trend that obviously will not lead to more accurate results.

Phase I
Base Year
Inventory

The first phase is the collection of any data that may be relevant to a description of how the transportation system behaves and why. It includes an inventory of

existing transportation facilities and their characteristics;

existing travel patterns, as determined through origin-destination (O-D) surveys (including data on choice of modes or routes) and traffic counts on the main transportation links; and

planning factors, such as land use, distribution of income, neighborhood structure, and types of employment.

It should also include the gathering of any available data relating to historical trends in the inventory; for example, growth in population, city size, and car ownership.

There are two basic purposes for this phase. First, one should make use of the fact that any future transportation system probably will be only an extension or modification of the present one: Some of the features of the present system will be included as part of the system at some future time and some of the patterns of travel may also remain the same. The second purpose is to obtain numerical values of certain parameters, which will appear in the mathematical models developed in phase II.

Phase II
Model Analysis

The purpose of this phase is to establish relations between various quantities measured in phase I. In effect, the data of phase I are reduced to formulas, some of which may be valid in the future. These relations (models) are usually classified in the following categories:

1. Trip Generation Models. These models attempt to relate

trip production and/or attraction (that is, the number of trips originating or destined for any region) to such things as land use (industrial, residential, business), population densities, income distribution, and type of employment.

2. Trip Distribution Models. Given the productions and attractions of various zones described in (1), the next step is to develop formulas that describe how trips from any origin are distributed among the various destinations; that is, to reduce the O-D pattern to some simple equations involving the productions and attractions of the zones and perhaps a few other quantities, such as the distance or cost of travel from an origin to a destination.

3. Traffic Assignment Models. Given the trip distribution (O-D table) described in (2), the last step is to determine how these trips distribute themselves over the various routes between each origin and destination (including choice of mode); one wishes to relate the flows on various routes to the physical characteristics of the transportation facilities (and the O-D table).

There are certain modifications of this classification of models. Some people consider the models for modal choice separately from the traffic assignment within the various modes. On the other hand, others have combined the distribution and assignment models into models relating the flows directly to the productions and attractions (and the properties of the transportation network).

Phase III
Travel Forecasts
(Extrapolations)

In this phase, some predictions are made of future land use, population, and so forth, based primarily on extrapolation of any relevant historical trends. The trip generation models developed in phase II are then used to estimate the resulting productions and attractions of trips, and the distribution models are used to estimate a future O-D table. Regardless of what special procedures are used, the goal is to extrapolate some known trends in certain observables so as to construct an estimate of future trip generation and distribution.

Phase IV
Network
Evaluation

Having produced an estimate of a future trip distribution in phase III, the final phase involves the assignment of these trips to routes on various transportation networks that include proposed new facilities. Comparisons of costs and benefits are then made between predicted flow patterns on the various networks. The ultimate purpose is to provide a basis for an economic evaluation of proposed new facilities.

Because the cost of travel between various points may influence the trip distribution and even the projected land use and trip generation, it may be necessary in evaluating some proposed network to recalculate the O-D table in phase III for each network. It may even be necessary to do some iterative calculations. One may need some preliminary estimates of travel costs in order to estimate the trip distribution in phase III. From this, one can calculate new estimates of travel costs in phase IV and use these to reevaluate the trip distribution and travel costs. It is for this reason that some people prefer to have models that combine the distribution and assignment. Phase III would be used to estimate only the trip generation, and phase IV would include the combined distribution and assignment.

When these procedures were developed in the 1960s, transportation planners expected a continued reliance on the private automobile as the primary mode of transportation in most American cities. Consequently, the extrapolations in phase III reflected this bias, and the evaluations in phase IV were oriented mostly toward expansion of the highway system. In the 1970s, however, the emphasis in planning (and politics) shifted toward an anticipated expansion of the public transportation network. There was great concern over such issues as pollution, the effect of highway expansion on the social aspects of urban life, and the projected scarcity of fuel for private transportation. Emphasis in the "network evaluation" phase consequently shifted toward prediction of the consequences of new transit lines, expanded bus service, personalized rapid transit systems, small vehicles, various demand-responsive public transportation systems (Dial-a-Ride), or car pools.

The main questions that have been generated by this change in direction are concerned with the modal split in the traffic assignment. Much of the effort in transportation planning in the 1970s has therefore been directed toward "choice models" in an attempt to predict how the attractiveness of a proposed mode depends on its physical characteristics (speed, comfort, frequency of service, access). It is obviously much more difficult, however, to predict the consequences of adding to the network facilities or modes with characteristics that are different from those in the existing system than to predict the consequences of simply expanding the existing system with facilities of the same type as already exist.

1.3 Critique

No one could seriously object to the basic logic of the four-phase transportation planning procedure if it were possible to follow it completely and accurately. In essence, the implication of the procedure is that by observing present patterns of travel and perhaps historical data of past patterns one can infer certain empirical relations between various observations. Although the values of the quantities in these relations may change with time, the relations themselves are presumably independent of time. Thus, among all variables that are relevant to transportation planning, the evolution of some of these can be inferred (through the empirical relations) from the evolution of others. If, for example, it were possible to determine the time evolution of such things as population, car ownership, and income and extrapolate these into the future, and if one could relate all other quantities of interest in transportation to them, plus properties of a proposed network, then one could also infer the time evolution of the latter.

Regardless of the pros and cons of various aspects of this outline, the fact remains that certain limited data about the historical evolution of certain quantities are at our disposal, and we have no possibility of collecting new data on past behaviors. We also have the option, at a price, of making studies of anything relevant to the future in the present behavior. From this, we want to make predictions of the future and the effect of various policies on the future. Ideally, we would like to do this as accurately as possible within the limitations of available data and estimate the probable errors, which, depending on the circumstances, may be larger than one would like them to be.

The main objections to the planning procedure concern the detailed manner in which it was actually carried out or the manner in which it evolved rather than with the abstract principle. The first people who devised and attempted the outline of steps showed remarkable imagination. It was certainly worth trying a few times to see how it would work on a small scale. Instead of using pilot studies to examine the weaknesses of the method and devise improvements in logic, most subsequent studies only became larger, with more data but not much improvement in accuracy or technique.

Some of the major deficiencies of the procedures, as executed in practice, are the following:

1. The final stage, evaluation, has received little attention. The planning projects based on this scheme often, conveniently, ran out of funds by the time they reached this phase. The project phases were, for the most part, performed sequentially; that is, the outline was not just a conceptual framework, it was a work schedule with the evaluation phase at the end.

It is rather difficult to make a sensible study without knowing what specific questions to answer. If one knows what things to analyze, then one might be able to design an experiment so as to collect the data that seem most relevant, collect a reasonable amount of data, and so forth. Unfortunately, the attitude of some people who promoted these studies was that one should assemble a "data bank." If there is enough data, all questions can be answered or, in any case, any data can be used for something. This may appear to be true, in principle, but is false in practice. One can easily assemble an arbitrarily large collection of irrelevant data. If it contains something useful, it may be just as expensive to retrieve from the computer tapes, where it is mixed up with all sorts of junk, as to perform a new experiment. People who must design experiments soon learn that they should not collect more data than is necessary.

The solution is not obvious. These are expensive experiments, but it is easier to convince a government agency to appropriate money for a well-defined project whose results one can see (a pile of data) than for a study of how to do a study.

2. These studies became popular very suddenly relative to the time scale in which cities show significant growth. At the time of the early studies, people were necessarily confronted with the problem: We can collect data today, but we have historical data on only certain things, such as records of population and land ownership and old photographs.

Any estimate of what will happen in the future involves extrapolation. The "laws" of physics have been observed over and over again in the past and have always agreed with the experiment. Everyone (except some prophets of doom) believes that they will be true throughout the foreseeable future. It is an extrapolation, however. What has been constant throughout the known past is likely to be constant in the future. For phenomena involving human behavior, however, we make extrapolations of a more risky type; for example, extrapolations of population growth.

From past experience we have learned something about reliability of population estimates, how well they fit exponential growth or other things. Extrapolations work well over short periods of time (except for plagues, wars, and so forth), not as well for long periods of time, but never with arbitrary precision. Other phenomena, such as the growth of air traffic or the decline of rail traffic, we also try to extrapolate, but with considerably less accuracy. In 1900 we could not possibly have predicted the present air traffic, but might have done reasonably well with population. Even within the time horizon of the studies made in the 1960s, transportation planners did not foresee the trends that were to occur in the 1970s: the virtual end to freeway construction, the effects of fuel shortages.

In a hypothetical situation in which one knows nothing about the past, one can only observe the present; extrapolation is quite arbitrary. Mathematically, it is like saying that if there exists a function $f(t)$ with a known value at $t = 0$, what is its value at time 10? To say that in the present time I observe some relation among quantities (such as a trip length distribution in a distribution model) is equivalent to saying that there is an equation $f(t, x_1, x_2, \ldots) = 0$, which is true at time 0 (x_j are some observed numbers). For $t \neq 0$, $f(t, x_1, x_2, \ldots)$ could be anything (maybe even 0). There is, of course, an even more extreme situation. One might conjecture that the relation $f(t, x_1, x_2, \ldots) = 0$ is true at $t = 0$, perform an experiment, find it does not work well, but still assume $f(t, x_1, x_2, \ldots) = 0$ for $t > 0$.

In transportation planning, one is trying to extrapolate time functions into the future for which there is no historical numerical data. The only thing that saves us from the ridiculous is that we do, in fact, have qualitative information. Perhaps we know that people do or do not live in the valley because it is hot there. Before computers, planners estimated city growth and land use from mental extrapolations of past experiences. People with good memories and good judgment could do this quite well. They could actually in some way call upon a memory (the brain) having a greater capacity than a computer and they could call upon a logic with three-dimensional visual patterns that is very difficult to code in a computer. Despite the fact that the "modern" methods of planning have been highly computerized, they do, at least indirectly, exploit some human judgments.

Unfortunately most studies are designed to be completed in a few years and then terminated; or in any case, they start with a very concentrated effort, to be followed by a less intense period of "updating." In fact, they should be done in reverse order. In the early stage, one knows very little and cannot plan very accurately. As time goes on, one can start to accumulate historical data, check the validity of tentative conclusions, and otherwise prepare to do a more sophisticated job in the future. Much of the "data" based on qualitative conjectures can at later stages be programmed with real data.

3. These studies have limited budgets, personnel, and so forth, partitioned among various tasks. To achieve maximum accuracy of the final conclusions, most of the money should go toward improving the accuracy of those aspects contributing the largest error to the final results. It is rather difficult to define precisely how this should be done, since errors are not known very well, but some qualitative estimates would help (even to within a factor of 10).

A significant fraction of the budget in most of these studies went into collection of such things as O-D data and home interviews. This is done so accurately that some gross features of the present traffic distribution are known within an error of a few percent or less. These data are then used to fit some "distribution model," which is likely to be incorrect by a factor of 2. This exercise of "calibration" also consumes considerable computer time, which takes another good part of the budget. The unfortunate fact is that most of the information contained in the original O-D data is, in effect, thrown away when it is used only to construct a model with four or five parameters.

Considerable effort also goes into the calculation of shortest travel times between various points in the network and the evaluation of possible flow assignments. Parts of this are done with great precision but are applied to a network that is only a crude representation of the real network, with a future O-D table that is only qualitatively correct (as is the conjecture that people choose the shortest route).

The criticism here is not so much with the crudeness of certain parts of the procedure; this may be impossible to correct. Rather, the criticism is with the wasting of so much effort on certain aspects of the procedure to achieve high accuracy that cannot be profitably used. Most large-scale studies would have been much more successful if they had operated on one tenth of their budgets.

1.4
Other Problems
The characteristics of a transportation system can be stratified roughly into various levels of detail. One can study the properties of individual vehicles and their interactions with each other, the interference between small systems (traffic intersection, terminal buildings), or properties of large systems. There are no well-defined boundaries. Any change in individual behaviors of vehicles influences the gross behavior and, to some extent, conversely. In the sense that a study of a large system includes the study of the detailed properties of its components, the study of transportation networks could be interpreted to include nearly all aspects of transportation, including the theory of traffic platoon motion. This is a little like saying that the theory of building construction includes thermodynamics, which includes atomic physics. In fact, most problems involving

large systems can be analyzed entirely in terms of the inter-relations between gross properties and almost completely divorced from the relation between microscopic and macroscopic properties.

Transportation planning procedures include many interesting types of mathematical problems, even if we limit the scope of the subject to the interrelations among gross properties. There are many problems in experimental design that have not been investigated (or at least not applied). Most of them are related to probability and statistics because the main questions deal with analysis of errors, reliability of extrapolation, and so forth. In what follows, we will avoid any topics involving stochastic analysis. We will be concerned mainly with mathematical representation of a network, characterization of flows on a network, and consequences of modification in a network. We also will avoid, for the most part, questions involving economics or social values.

The purpose of this brief commentary on transportation planning is to emphasize that in the study of networks the goal should be to achieve maximum simplicity and appre-ciation of important issues rather than high accuracy of numerical results because even those things that could, in principle, be done accurately will be using inaccurate data.

References There exists a vast literature on the transportation planning process, including published reports for most metropolitan regions to which the procedures have been applied. The following is a list of a few recent surveys or textbooks.

1
Creighton, R. L. *Urban Transportation Planning*. Urbana, Illinois: University of Illinois Press, 1970.

2
Catanese, A. J., ed. *New Perspectives in Urban Transportation Research*. Lexington, Mass.: D. C. Heath & Co., 1972. An interesting view expressed in this book (although I do not endorse it) reads, "Presently, the transportation planning process is fairly well-defined and its acceptance is, for the most part, taken for granted."

3
Urban Traffic Models, Possibilities for Simplication. Paris: Organization for Economic Cooperation and Development, 1974.

4
Stopher, P. R., and Meyberg, A. H. *Urban Transportation Modelling and Planning*. Lexington, Mass.: Lexington Books, 1975.

5
Domencich, T. A., and McFadden, D. *Urban Travel Demand: A Behavioral Analysis*. New York: North Holland Publishing Co., 1975.

2 MATHEMATICAL ABSTRACTIONS

2.1
Spaces

In the analysis of a transportation system, or any other physical system, it is desirable to start by extracting only a few of its properties, analyzing these, and then adding more and more structure as is necessary to determine the answers to relevant questions. One should not describe the system in any more detail than necessary. The analysis of a complicated system can be difficult enough without clouding it with irrelevant information.

One of the most primitive of all mathematical concepts is that of a space. A *space* is a collection of objects, called *points*, identified by some labels. A *finite space* is a collection of a finite number of objects. As yet, this space has no properties, no distance, and no operations, such as addition. The notion is so primitive that about all one can do in comparing one space with another is to see if they have the same number of points (or if one space has more than the other) in the sense that there is a one-to-one correspondence between the points of the two spaces. The notion that a certain collection of objects can be compared with another collection of objects is very basic to the principle of counting, which we all learn at a young age. Before one can master the art of counting, one must first recognize that collections such as three apples and three bananas have something in common, namely they can be paired off with each other.

We may designate a space by a single symbol S or by listing the points in the form {label one, label two, . . .} or {1, 2, . . .}. Regardless of what physical properties may be identified or associated with the labels (for example, a set of apples are good to eat), the recognition that these objects form a space associates with these objects *only* the property that they can be labeled and compared in number with another collection of objects. Furthermore, two spaces with the same number of points are considered equivalent regardless of what the labels may be. Here are some examples:

1. The integers $\{1, 2, \ldots, n\}$; a collection of n highway intersections {Main street–1st street, Main street–2nd street, ..., Main street–nth street}; a collection of n memory locations in a computer; a collection of n locations in a sheet of paper. (It is, of course, basic to the notion of mathematics or symbolic logic that all these have something in common.)

2. The real numbers between 0 and L, that is, the set of numbers x such that $0 < x < L$, designated in usual mathematics notation as $\{x \mid 0 < x < L\}$; the set of possible locations of a car on a highway of length L (miles, centimeters, or whatever).

3. The set of ordered pairs of real numbers $\{[x, y] \mid -\infty < x, y < \infty\}$; the coordinates of points on a sheet of paper; the positions of points on the earth (idealized to be flat and infinite).

A *subspace* S' of a space S is a collection of points x, each of which is also in S. The expression $S' \subset S$ means S' is contained in S; that is, every point of S' is in S. If $S' \subset S$ and $S \subset S'$, then S and S' are equivalent, that is, $S = S'$. For example, the integers $\{1, 2, \ldots, n\}$ form a subset of the real numbers; intersections form a subset of the points on a highway system, which form a subset of the points on the surface of the earth.

In the mathematics literature there are entire fields of specialization devoted to "set theory" and "number theory." It is not the purpose here to give even a hint about what these fields contain but rather to emphasize that physical problems of widely different natures can abstractly be identical or at least have some features in common. As the complexity of the mathematics and its associated applications increase, the similarity between problems of different physical context becomes less obvious. The test of a good applied mathematician (or even an engineer) is that, in seeking the answer to some question, he can separate the relevant features from the irrelevant features, recognize the nature of the problem, and exploit its similarity with other problems.

2.2 Graphs   To an engineer a "graph" is a curve drawn on a sheet of paper to represent a function. To an abstract mathematician a graph is something quite different; it is a formal generalization of the notion of a collection of points (in a plane, perhaps) connected by some (directed) lines.

A (directed) *graph* G is a space N of points called *nodes* plus a space L of points called links. The links are ordered pairs of nodes in N. The graph G is denoted by the symbol $\{N; L\}$.

A graph is itself a "space," in the sense that it is a collection of labeled objects, but it happens to consist of two different types of objects. The first types of objects, the nodes, have labels and so form a space (a subspace of the previous space). If there is a finite number of nodes, we can label them $\{1, 2, \ldots, n\}$. The second type is ordered pairs of objects from N. If the nodes are labeled $i, j, k, \ldots$, then the ordered pair (i, j) is called a link from i to j. The space L need not contain all possible ordered pairs, but it could contain a link (i, i) from i to i. If (i, j) is in L, symbolized $(i, j) \in L$, then for $i \neq j$ the symbol (j, i) is considered different from (i, j) and may or may not be in L.

A graph G is defined by a listing (in any order) of all the node labels plus a listing of all the link labels. An expanded notation for G might be

$$G: \{\alpha, \beta, \gamma; (\alpha, \beta), (\beta, \gamma)\},$$

designating that there are three nodes with labels α, β, and γ, and two links with labels (α, β) and (β, γ).

The relevance of graphs to transportation arises from the fact that road or airline maps, for example, can be considered as graphs. The nodes might be street intersections or air terminals, and the links might be directed street connections between intersections or directed air routes between terminals, represented on the map by lines with direction arrows between pairs of points. If nodes i and j are in fact connected by a two-way route, this route can be represented as two links, (i, j) and (j, i). Physically, a two-way road consists of two adjacent one-directional lanes, although on a map it would usually be represented by a line with no direction arrows rather than by a line with arrows in both directions.

There is a danger in trying to identify an abstract notion too closely with a particular type of mental picture. Pictures usually embody elements of structure that are not necessary to the abstract graph. On the other hand, an abstract graph may include elements that cannot be easily embodied in pictures. For example, the definition of an abstract graph does not involve the notions of distance between points or angles between links, but these notions are embodied in any two-dimensional picture and may distort one's intuition about some properties associated with the graph. If one tries to draw a picture of a graph with many links, one may find that it is not possible to draw the links so that they do not cross. Also, a link from i to i may have no physical interpretation. The abstract graph, however, is perfectly well defined regardless of whether one can draw a two-dimensional road network from it.

A graph is a rather primitive mathematical concept; it does not have many of the properties of the physical system from which it is abstracted. A set of nodes can be any collection of things, such as points in space, sections of highways, numbers, or pieces of paper; the links need not have any physical meaning. Given a collection of n nodes, there are n^2 possible ordered pairs. A graph is specified if one identifies which of the n^2 possible ordered pairs are actually in L and, therefore, which ones are not in L. Thus a graph of n points is a collection of n^2 yes or no answers to some question relating to pairs of nodes. Does an ordered pair of nodes (i, j) have some property that will identify it as belonging to L, or does it not? In transportation, the property in question is usually: Is there a physical path directly from i to j without passing through other nodes? Or the property might be: Is there a path of any sort from i to j?

An application of graph theory that touches nearly all scientific fields is the following. We take as the set of nodes (objects) the measurable properties of a system (such as the pressure, temperature, and density of a gas); each node is a label specifying a property. A link represents an ordered pair of properties. We can then use a graph to identify the

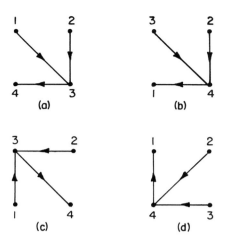

Figure 2.1
Examples of equivalent and/or
identical graphs.

interrelations among properties; that is, $(i, j) \in L$ if a change
in property i induces a change in property j. From an
analysis of the graph, one would presumably discover
whether there were some set of properties unrelated to others.
(This illustrates the primitive nature of the subject, however,
in that it only deals with yes or no answers, not "how much"
or "how far.")

**2.3
Representation
of Graphs**

An abstract space has no "structure." One space of n objects
is equivalent to any other space of n objects. It is a trivial
exercise to decide if one space is equivalent to (or larger or
smaller than) another, at least if the space is finite. This is not
true of graphs.

To represent a geometric figure of points and lines as a
graph, we would first label all the points (with numbers
perhaps) and then make some listing of the lines as links with
labels (i, j). The figure itself would not be changed if we were
to number the nodes in some other order or change the order
of listing the nodes and links. Thus, using the notation
$\{N; L\}$ for graphs, there might be a number of graphs that
we could in some sense consider equivalent. For example,
the graph of figure 2.1a might be considered equivalent to

that of 2.1b because it differs only in a renumbering of the nodes; but 2.1a has the representation $\{1, 2, 3, 4; (1, 3), (2, 3), (3, 4)\}$, or any other ordering of these nodes and links such as $\{4, 3, 2, 1; (3, 4), (2, 3), (1, 3)\}$, whereas 2.1b has the representation $\{1, 2, 3, 4; (3, 4), (2, 4) (4, 1)\}$.

The graph of figure 2.1c, on the other hand, has exactly the same notation as that of 2.1a, and 2.1d has the same notation as 2.1b; that is, their graphs are the same. The nodes and links in 2.1c have been rearranged on the figure in a different geometry from 2.1a, but this does not affect the properties of the graph. It follows that the graph of 2.1a is equivalent to that of 2.1d, and 2.1b is equivalent to 2.1c.

Generally we define two graphs $G = \{N; L\}$ and $G' = \{N'; L'\}$ as being *equivalent* if it is possible to relabel the nodes of N, thereby also changing the labels of L, in such a way that the set N is mapped into N' and L into L'. Formally G and G' are equivalent if for every $a \in N$ there is a node $g_a \in N'$ and, conversely, for every $g \in N'$ there is a node $a_g \in N$; that is, g_a is a one-to-one mapping. Furthermore, the set of links L' is the set of (g_a, g_b) with $(a, b) \in L$.

If we number the n nodes of G and G', then the mapping g_a is a permutation of the numbers $1, 2, \ldots, n$. There are $n!$ possible permutations, thus potentially as many as $n!$ equivalent graphs, although many of these $n!$ mappings may lead to exactly the same set of nodes and links; that is, $N = N', L = L'$. Although, in principle, it is straightforward to list all the graphs $\{N'; L'\}$ that are equivalent to a given graph $\{N; L\}$ and consequently see if two graphs are equivalent, this is certainly not easy to do in practice.

Much of abstract graph theory deals with the classification of graphs into equivalence classes and with comparisons of their structural properties. Little of this theory is relevant to transportation, however. There are many possible ways of translating a figure of points and directed lines into a graph by labeling the nodes in different ways, such as in figures 2.1a and 2.1b. However, once a labeling of the nodes has been decided, as in figure 2.1a, it is generally not particularly relevant to transportation questions that another collection of links on the same set of nodes leads to an equivalent graph; for example, it is not relevant that figures 2.1a and 2.1d are equivalent graphs. The equivalent graphs do not describe the same physical picture.

Hereafter, we will assume that a scheme of labeling the nodes has been established. We will consider two graphs as being *equal*, $\{N; L\} = \{N'; L'\}$, only if the set N is the same as N' and L is the same as L', although the nodes and links of N' and L' need not be listed in the same order as N and L. Thus figures 2.1a and 2.1d are not considered to be equal, or if one decides to number the nodes as in 2.1b, then 2.1b and 2.1c are not equal.

There are many possible ways (notations) of representing or identifying a particular graph. Listing all the nodes and all the links is one possibility, well suited to computers, which prefer sequential logic or enumerations, but perhaps not the best for quick visual identification or manipulation. Another possible notation is the matrix (two-dimensional array). If one has a graph of n nodes numbered according to some specified pattern, then the links are identified by noting whether each symbol (i, j) is in L. The matrix is a common way of representing a function on such index pairs. In the present case the function is yes or no, zero or one, or any other two-valued function. A graph on 5 nodes, for example, can be identified by a 5×5 matrix:

$$
\begin{array}{c c}
 & \begin{array}{c c c c c} & & j & & \\ 1 & 2 & 3 & 4 & 5 \end{array} \\
\begin{array}{c} \\ \\ i \\ \\ \end{array}
\begin{array}{c} 1 \\ 2 \\ 3 \\ 4 \\ 5 \end{array}
&
\left|\begin{array}{c c c c c}
0 & 1 & 0 & 0 & 0 \\
0 & 0 & 1 & 0 & 0 \\
1 & 0 & 0 & 1 & 0 \\
1 & 1 & 0 & 1 & 0 \\
0 & 0 & 0 & 1 & 1
\end{array}\right.
\end{array}
$$

Each entry in this array can be identified by coordinates (i, j). If the symbol (i, j) is in L, we put a 1 in the (i, j) block; if not, we put a 0 there. Thus a graph G on nodes $N = \{1, 2, \ldots, n\}$ can be represented as a matrix:

$$
a_{ij} = \begin{cases} 1 \text{ if } (i, j) \in L, \\ 0 \text{ if } (i, j) \notin L, \end{cases} \qquad 1 \leq i, j \leq n.
$$

Certain special classes of graphs have easily recognizable pictures. Three of these are the following:

1. An *undirected graph* is one in which all links are two-way; that is, if $(i, j) \in L$, then $(j, i) \in L$ for all i and j. The matrix of such a graph is said to be *symmetric*. More generally, any matrix of numbers a_{ij} is symmetric if $a_{ij} = a_{ji}$ for all i and j. In the present case the numbers a_{ij} are either 0 or 1. Geometrically the links (i, i) lie along the diagonal from upper left to lower right. If a matrix is reflected over this diagonal, rows map into columns, so that location (i, j) maps into (j, i); only the diagonal locations remain fixed. A graph is undirected if this reflection maps the matrix into itself. In an undirected graph, the matrix elements on one side of the diagonal uniquely determine those on the other side; thus it suffices to show only the diagonal elements and those on one side.

2. A *complete graph* has a link between every pair of nodes, its matrix has all 1s. This is mathematically trivial, but it arises in many applications; for example, if i numbers some trip origins or destinations, the graph might represent whether anyone wishes to go from i to j or whether it is possible to go from i to j by some route.

3. Suppose all nodes in graph G can be classified into a set of nodes X and its complement $\bar{X}$ (those nodes not in X), and L consists of all ordered pairs joining nodes in X to nodes in $\bar{X}$. This type of graph arises in the analysis of shipments of goods from suppliers to consumers, provided no node represents both a supplier and a consumer. If the nodes of X are numbered first, then those of $\bar{X}$, the matrix of L has the form

	X	$\bar{X}$
X	0	1 1 ... 1 1 ⋮ ⋱ 1 1 ... 1
$\bar{X}$	0	0

where the 0s stands for null submatrices of all zeros. Such a graph is called a *bipartite graph*. (Note that if one were given a matrix for a bipartite graph in which the numbering was all mixed, one would not recognize it as such by inspection.)

2.4 Routes Suppose we have a graph representing a transportation network with nodes labeling junctions, origins, or terminals and links describing pairs of nodes that have "direct connections" (for example, nonstop flights or roads that connect the nodes without going through other nodes).

About the only practical questions that can be answered using only graph theoretical arguments are those relating to whether it is possible to reach one node from another by traveling either directly or via other nodes. Although this may at first seem like a small matter, it arises over and over again in the analysis of flows on networks, not so much with the actual transportation network as with subgraphs. Particularly in the analysis of congestion, one is often confronted with the question of what will happen when certain links have reached capacity and are therefore not links in the graph seen by new travelers.

The terminology in graph theory can get somewhat confused because, for certain purposes, it is desirable to know whether it is possible to go from one place to another even if this means traveling in the wrong direction on a directed link. The terms route, path, or chain are therefore used to denote different things by different writers. Here we will have only one definition of a means of going from one node to another; we might refer to it as a route, chain, or path.

A *route* (path, chain) from node n_1 to node n_r is an ordered sequence of links $(n_1, n_2), (n_2, n_3), \ldots, (n_{r-1}, n_r)$ with $n_j \neq n_k$ for any j and k, and $(n_i, n_{i+1}) \in L$. A *cycle* is an ordered sequence of links $(n_1, n_2), (n_2, n_3), \ldots, (n_{r-1}, n_1)$ with $n_j \neq n_k$, and $(n_i, n_{i+1}) \in L, (n_{r-1}, n_1) \in L$. The "physical" meanings of these definitions are obvious. If there is a link (n_1, n_2) on a transportation network, then one can go from n_1 to n_2. If there is also a link (n_2, n_3), then one can go from n_1 to n_3 in two steps, even if there is no link (n_1, n_3). The route is a sequence of links that will take one from n_1 to n_r without passing through any node more than once. A cycle is a "closed route"; it takes one from n_1 to n_1 without passing through any node more than once.

A more general notion of a sequence $(n_1, n_2), (n_2, n_3), \ldots, (n_{r-1}, n_r)$ without the restriction $n_j \neq n_k$ is possible but not very useful. It is a trivial generalization because any such sequence can be decomposed into a route from n_1 to n_r plus some cycles, the important point being that no point n_r can be reached by this more general sequence that cannot also be reached by a route. The decomposition is a geometrically obvious one. We start out from node n_1 and proceed to $n_2, n_3, \ldots$, until we encounter for the first time an n_k with $n_k = n_i$, $i < k$; that is, we have returned to a previous node. The sequence $(n_i, n_{i+1}), \ldots, (n_{k-1}, n_k)$ is a cycle because $n_k = n_i$, and because this is the first such occurrence, all other $n_j, j < k$, are different. Now remove this cycle from the original sequence and consider the sequence $\ldots, (n_{i-1}, n_i), (n_k, n_{k+1}), \ldots$. We now remove the first cycle from this sequence (if there is one) and continue by induction until we have constructed a sequence from n_1 to n_r with no cycles.

We will say that node i is *accessible from j* if there is a route from j to i. A graph will be called *completely connected* if there is a route from every node to every other node.

There is a danger in having a terminology that is too closely identified with physical pictures. In transportation we usually think of nodes as being intersections, interchanges, terminals, and so forth, and routes in the mathematical sense as being routes in the geometric sense. The nodes, however, could be roads, air connections, or any other labeled objects, and the links could be the *possibility* of making a transfer from one node to another (at any unspecified physical location). The danger comes from an oversimplification of a physical situation. Intersections, terminals, and so forth may have some internal logical structure. At an intersection one may not be able to make left turns. At a terminal one may not be able to transfer between certain physical routes that "meet" there. One of the subtle points contained in our definition of a route is the implication that nodes have no internal structure: one can go from n_1 to n_2 on a link and then from n_2 of one link to n_2 of another link, and from there to n_3.

Given a matrix representation of a graph, one can test whether a sequence $(n_1, n_2), (n_2, n_3), \ldots, (n_{r-1}, n_r)$ with $n_j \neq n_k$ is a route by checking whether each (n_k, n_{k+1}) is in L. For example, start with the n_1th row and see if there is a 1 in the n_2th column. Then look at the n_2th row and see if there is a 1 in the n_3th column, and so forth.

2.5
Accessibility Graph

Because a graph can be interpreted as a set of yes or no answers to some question associated with all pairs of nodes, it must be possible to associate with any graph $G = \{N; L\}$ another graph $G_a = \{N; L_a\}$ having the property that $(n_i, n_j) \in L_a$ if n_j is accessible from n_i in the graph G.

In most practical applications, the graph G_a is the complete graph (if G is completely connected); but if it is not, one should know about it. It is sufficiently easy to construct G_a and analyze its structure; this should always be done immediately with any graph, if only to check that graph G has only the properties it was supposed to have.

An algorithm for either hand or computer construction of G_a might go as follows. To construct row 1 of the matrix for G_a, first insert 1s wherever they exist in G (any node that has a link from node 1 is accessible from node 1). Second, observe those node numbers j with $(1, j) \in L$ and select the corresponding jth rows of G. Insert 1s in any new locations in row 1 of G_a where there are 1s in any of these rows of G. The first row of G_a now has a 1 in any column that can be reached in either one or two steps from node 1. Observe also which new node numbers were added in this last step, select the corresponding rows of G, and add 1s to row 1 of G_a in any new locations where there are 1s in any of these rows. There are now 1s in any column of row 1 that can be reached in one, two, or three steps from node 1. Continue this until at some iteration no new 1s are added to row 1 of G_a. We have now found all nodes accessible from node 1, and we can fill the remainder of row 1 of G_a with 0s.

The other rows of G_a can be constructed in the same way, but it may be faster to replace G first by a matrix consisting of the completed first row of G_a and the remaining rows of G. If, in constructing any other (second) row of G_a, one should find that node 1 is accessible from the second node, then *all* nodes accessible from node 1 should be added to the second row. Similarly, after completing a second row of G_a, it should be inserted into G for the evaluation of a third row, and so forth. After a few rows of G_a have become nearly filled with 1s (if this in fact happens), this procedure will go rather quickly. If the rows do not get filled with 1s and the matrix is not too large, one should at least begin to see a pattern emerging. To further accelerate the procedure, it may be advantageous, after completing row 1, to choose a row known to have a link to node 1.

If the graph G is not completely connected, then it may be advantageous to reorder the nodes (or construct an equivalent graph) so as to display its structure in a more convenient way.

2.6 Decomposition of a Graph

From a graph G_a (or its matrix) one can see immediately which pairs of nodes i and j have the property that both i is accessible from j and j is accessible from i: graph G_a has a 1 in both the (i, j) and (j, i) locations. We will now use this property to partition the nodes N into subsets.

We first form a set of nodes N_1 consisting of node 1 plus all nodes j accessible from 1 and from which 1 is accessible. This set N_1 has the further property that every node of N_1 is accessible from every other node: Node 1 is accessible from any node $j \in N_1$ and so, therefore, is any node that is accessible from 1, namely all nodes of N_1.

Now we pick any node i not in N_1 and form a set N_2 of nodes accessible from and to i in the same sense. These nodes also have the property that every node of N_2 is accessible from every other node in N_2. The sets N_1 and N_2 are disjoint, for if any node of N_2 were in N_1, all nodes of N_2 would be in N_1 (which contradicts the hypothesis that at least one node i is not in N_1). We continue to form other sets $N_3, \ldots, N_m$ until N is exhausted.

If the listing of the nodes of N is reordered in such a way that nodes of N_j are listed consecutively (for example, first the n_1 nodes of N_1, then the n_2 nodes of N_2, and so forth), the matrix of G_a with the rows and columns written in the order of this listing will show a very special form. In this $n \times n$ matrix, one can identify an $n_j \times n_k$ block describing the accessibility of nodes in N_k from nodes in N_j. This block will either contain all 0s or all 1s because if any node of N_k is accessible from any node of N_j, all nodes of N_k are accessible from all nodes of N_j. In particular, the $n_j \times n_j$ matrix of nodes in N_j will contain all 1s.

From the point of view of accessibility, all nodes of N_j are equivalent. They are all mutually accessible, and any relation with nodes outside N_j that is true of one node is true of all. At this point we should not think of N_j as a *collection* of objects (nodes) but as an object itself; it is a single node in another graph G_a^* that has a link (N_j, N_k) if, in the original graph, the nodes of N_k are accessible from N_j. This graph can also be represented by an $m \times m$ matrix (hopefully of small dimension compared with the original graph) of 0s and 1s.

In order to decompose and reorder the listing of nodes in graph G_a^*, we first partition the nodes into subsets M_1, M_2, ... Each M_i is to be a minimal set of nodes with the following property: If node N_j is in M_i, so is any node N_k for which *either* N_j is accessible from N_k or vice versa. In effect we are (temporarily) converting the directed graph G_a^* into an undirected graph by considering each link as being two-directional. We are then constructing the accessibility graph of this undirected graph and partitioning the nodes in the same way as we obtain the nodes N_j from the original nodes. We would obtain the same sets M_i if we were to convert all links of the original graph into two-way links and find the Ns for this undirected graph.

If there is more than one M_i, then no node N_j of M_i (or original nodes in N_j, which is in M_i) is accessible from a node N_k of another M_l, or vice versa. The graph is like a street system on a collection of islands with no bridges.

Regardless of how the N_j may have been listed before, it is now advantageous to list them so that the nodes of M_1 (and the original nodes therein) are listed consecutively, followed by those in M_2, and so forth. In this new ordering of nodes, the matrix G_a^* has the "block diagonal" form

$$
\begin{array}{c|ccc}
 & M_1 & M_2 & \cdots \\
\hline
M_1 & A_1 & 0 & 0 \\
M_2 & 0 & A_2 & 0 \\
\vdots & 0 & 0 & \ddots
\end{array}
$$

in which A_1 is a matrix of 0s and 1s for links between the nodes of $N_j \in M_1$, A_2 is a matrix of links between the nodes $N_j \in M_2$, and so forth. The 0s represent matrices of all 0s between nodes of M_i and M_j, $i \neq j$. This structure applies not only to G_a^* but to G_a and the original graph G if all nodes are listed in the order described.

If we had known at the start that G would decompose in this way, we probably would not have considered this as a single graph. We would have constructed separate graphs for each of the completely unrelated pieces M_i. We would not have made a graph artificially more complicated by noting that the union of two disjoint graphs was also a graph.

In the second stage of the decomposition, we look at the nodes N_j within a single M_i or equivalently just assume that G_a^* has no such decomposition. We also go back to the directed graph G_a^* and observe some of its special properties.

If there is a link (N_j, N_k) in G_a^*, there must *not* be a link (N_k, N_j), $j \neq k$. If there were, the nodes of N_k would have been included in N_j, or vice versa. There is not even a route back to N_j. Every sequence of links from N_j must, therefore, eventually lead to a "trap," a node from which there are no exits.

One can quickly identify these traps from the matrix of G_a^*. They are associated with rows of G_a^* containing only a single 1 along the diagonal. (The corresponding column of G_a^* will contain 1s off the diagonal, for if a node has neither a way in nor a way out, it would have been eliminated in the previous step.) The traps are like the ends of dead-end, one-way streets.

In reordering the nodes N_j, we will list all traps (and the original nodes contained therein) first. If we were to erase from the original graph G_a^* these trap nodes and their associated rows and columns, the remaining matrix would also represent a graph with the same irreversible properties as G_a^*; it would also have traps, which, however, must have exits to the original traps (otherwise they would have been among the original traps). Once we have listed these nodes, we can strike them from the graph and look for new traps.

Having listed the nodes of G_a^* in this way, the new matrix G_a^* has the form

$\begin{matrix} 1 & & 0 \\ & 1 & \\ & & \ddots \\ 0 & & 1 \end{matrix}$	0	0	$\cdot$
A_{21}	$\begin{matrix} 1 & & 0 \\ & 1 & \\ & & \ddots \\ 0 & & 1 \end{matrix}$	0	$\cdot$
A_{31}	A_{32}	$\begin{matrix} 1 & & 0 \\ & 1 & \\ & & \ddots \\ 0 & & 1 \end{matrix}$	$\cdot$
$\cdot$	$\cdot$	$\cdot$	$\cdot$

The first block of nodes represents the first traps, the second block the second traps, and so forth. The matrix as a whole has a "block triangular" form with null matrices above the diagonal. The diagonal block matrices are the "identity" matrices. The A_{ij} are unspecified matrices, of which at least $A_{j,j-1}$ must be nonnull.

Again this structure of G_a^* implies a similar structure in G_a, which, in turn, implies a similar structure in the original graph G. For example, if in the bipartite graphs we had listed the destination nodes first, and also inserted links from i to i, we would have a matrix

$\begin{matrix} 1 & & 0 \\ & 1 & \\ & & \ddots \\ 0 & & 1 \end{matrix}$	0
$\begin{matrix} 1 & & 1 \\ & 1 & \\ & & \ddots \\ 1 & & 1 \end{matrix}$	$\begin{matrix} 1 & & 0 \\ & 1 & \\ & & \ddots \\ 0 & & 1 \end{matrix}$

2.7
An Application
A problem that occurs very frequently in applied mathematics (including some transportation-related problems) is solving n simultaneous equations for n unknowns, that is, determining real numbers $x_1, x_2, \ldots, x_n$ such that

$$g_i(x_1, x_2, \ldots, x_n) = 0 \qquad i = 1, 2, \ldots, n. \tag{2.1}$$

It is often true that each of the functions $g_i(\cdot)$ actually depends on only a few of the x_j. For example, if the equations are linear, the matrix of coefficients has mostly zero matrix elements. As a preliminary step in the solution of these equations, one may wish to see if there are some subsets of the system of equations that can be used to determine some of the unknowns. The question is, can one decompose this n-dimensional problem into problems of smaller dimension? The answer, unfortunately, is usually no, but if the problem can be decomposed into simpler problems, one would certainly like to know about it.

The indexing of the equations is arbitrary. If they have a solution, it must be true that each x_j appears in at least one of them. Furthermore, it must be possible to number the equations in such a way that the ith equation contains the variable x_i. Quite possibly the equations are actually given in the form

$$x_i = g_i^*(x_1, \ldots, x_j, \ldots, x_n; j \neq i). \tag{2.2}$$

To analyze the structure of these dependencies, we can construct a graph G with nodes $1, 2, \ldots, n$, and links L, $(i, j) \in L$, if the ith equation contains the variable x_j.

Suppose now we decompose the graph G and find that it is not completely connected; that is, the accessibility graph G_a is not complete. We proceed to reorder the listing of nodes (the indices of x_i) to separate any subsets of nodes M_1, $M_2, \ldots$. If there is such a decomposition, then clearly (2.2) can be written as separate sets of simultaneous equations for the x_i belonging to M_1, those belonging to M_2, and so forth.

We next rearrange the listing of the nodes within the sets $M_1, M_2, \ldots$ so that subgraphs are represented in a block triangular form. If, in particular, the equations formed from

(2.2) are linear, the matrix of coefficients will have the same block triangular form.

One can now see that the equations associated with the trap nodes N_1 contain only the variables x_i, $i \in N_1$. Although these x_i may appear in other equations (there may be links from N_j to N_1), the first n_1 equations represent a system in n_1 unknowns. If one can solve these equations for x_1, x_2, ..., x_{n_1}, then these known values can be substituted into the other equations. If one does this for all the trap nodes, then the equations for the second-order traps become sets of simultaneous equations involving only the variables in the sets N_j.

The whole problem can be reduced to solving a sequence of sets of simultaneous equations of dimensions n_1, n_2, ... Although one seldom finds that G is not completely connected, this exercise of decomposition is so simple that it is always worth trying.

2.8
Subgraphs and Compressed Graphs

A *subgraph* $\{N'; L'\}$ of a graph $\{N; L\}$ is a collection of nodes and links such that $N' \subset N$ and $L' \subset L$ (N' is a subset of N and L' is a subset of L), and $\{N'; L'\}$ is a graph (L' is a collection of links between nodes of N'). Subgraphs are very important in the study of transportation networks, first, because one can often use special types of subgraphs to help analyze features of the original graph; and second, because the graph one uses to represent a transportation network is usually itself a subgraph of a hypothetical "realistic" graph that contains more nodes and links than one can handle.

A *tree* is a graph (usually a subgraph of some other graph) that has one and only one route from a single node (called the "home node" or "root") to every other node. Geometrically (if one draws it in a certain way) this really looks like a tree with hierarchies of branches all having a path from the root. Its relevance to transportation arises from the fact that one often wishes to know the shortest or cheapest or fastest path from one node to all other nodes. The collection of links that are used in these extreme routes form a tree.

According to the scheme of listing described in section 2.6,

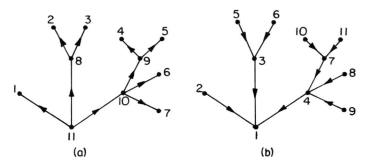

Figure 2.2
Two schemes for numbering the
modes of a tree.

the first collection of nodes are the tips of the branches,
numbered in any order. The next set of nodes is obtained by
cutting off the links to the branch tips and listing the new
tips, in any order. The root will be the last node listed.

An alternate scheme of listing nodes can be generated by
reversing the direction of all links so that all routes go from
the branch tips to the root. The root will now be listed first
(according to the ordering system of section 2.6), followed by
all nodes with direct links to the root, and so forth. This
second scheme will not, generally, be simply a reverse order
of the first. Figure 2.2a shows an example of a tree with the
nodes numbered in the order of the listing described in
section 2.6; figure 2.2b shows the corresponding graph with
all links reversed and renumbered. The matrix representations of these two graphs follow:

	1	2	3	4	5	6	7	8	9	10	11
1											
2											
3											
4				0							
5											
6											
7											
8	0	1	1	0	0	0	0		0		
9	0	0	0	1	1	0	0				
10	0	0	0	0	0	1	1	0	1	0	
11	1	0	0	0	0	0	0	1	0	1	0

	1	2	3	4	5	6	7	8	9	10	11
1	0										
2	1										
3	1	0									
4	1										
5		0	1	0							
6		0	1	0							
7	0	0	0	1		0					
8		0	0	1							
9		0	0	1							
10	0		0		0	0	1	0	0	0	
11					0	0	1	0	0		

These matrices do not have 1s on the main diagonal and, consequently, do not have quite the same form as the matrix G_a of section 2.6. If an N_j contained more than one of the original nodes, each node in N_j would automatically be accessible from itself and the G_a^* would, therefore, have 1s on the diagonal. If, however, an N_j contained only one node, and the original graph did not list a link (N_j, N_j), then the accessibility graph would have 0s on the diagonal unless we defined the accessibility graph in such a way that a node is always accessible from itself. For our purposes it is irrelevant whether a node has a link to itself; it is certainly implied that a node is accessible from itself.

If a set of nodes N of a graph $\{N; L\}$ is partitioned into two subsets X and $\bar{X}$, then the set of links

$$(X, \bar{X}) \equiv \{(i, j) | i \in \bar{X}, j \in \bar{X}, (i, j) \in L\}$$

is called a *cut-set*. The graph $\{N; (X, \bar{X})\}$ is a subgraph of $\{N; L\}$ containing only links between nodes of X and $\bar{X}$.

In transportation, the set of nodes X may be the nodes in some geographical zone. The cut-set is composed of the links (roads, perhaps) leading out of this zone. If one wished to make a survey of the traffic flow out of a zone with nodes X or investigate the capacity of the links out of this zone, one would need to look at all links in the cut-set $(X, \bar{X})$.

A *compressed graph* of a graph $\{N; L\}$ is a graph $\{N'; L'\}$ such that each node of N' is a subset of N (not necessarily a single node) and the set of N' is a partition of N; that is, each node of N is in one and only one of the nodes N_j of N'; and there is a link (N_j, N_k) in L' if and only if one of the original nodes of N_j has a link in L with some node of N_k.

An example of a compressed graph appeared in section 2.6, where the N_j were completely connected subsets of N and all nodes of N_j were considered to be equivalent. If, for a tree, we were to define N_j as the set of nodes $j - 1$ steps from the home node, the graph $\{N'; L'\}$ would have a very simple structure; L' would contain only the links (N_j, N_{j+1}). If, in the graph $\{N; (X, \bar{X})\}$, we let X and $\bar{X}$ be the nodes of N', $\{N'; L'\}$ would have just one link.

A subgraph or a compressed graph has fewer nodes and/or

links than the graph from which it is constructed and should, therefore, be easier to analyze than the original graph. Most real transportation networks contain more structure than can be analyzed even with a large computer, but this is usually not a problem because no one would use the details of such an analysis anyway. In analyzing a network one should always approximate it by the crudest graph that will still allow one to determine answers to whatever questions may be posed. We have not said what the questions are, nor is it obvious in what sense one graph is an approximation of another (the two problems are clearly related), but the main techniques for simplifying a graph are to use subgraphs, compressed graphs, or a combination of the two.

From the purely graph-theoretical point of view, we have no means of measuring how nearly one graph resembles another. We know when one graph is the *same* as another or equivalent to another, but not how "near" they are. There are a few things, however, that, as a practical matter, one should try to avoid. If one replaces a graph by a subgraph, there is a danger that the original graph may be completely connected but not the subgraph (a critical link is left out). In replacing a graph by a compressed graph, one runs the opposite risk: If the original graph was not connected (two islands, for example), the compressed graph may be connected. The implication in the analysis of a graph is that the node itself is internally connected. If, in compressing a graph, one (accidentally) compresses two nodes that originally were not accessible from each other, one will create an artificial link (a nonexistent bridge between islands) between disconnected subgraphs.

2.9 Simplified Transportation Networks

In a transportation study of a metropolitan region, the real transportation netwrok is always replaced by an approximating network containing a manageable number of nodes and links. The region is first divided into geographical zones of various sizes depending on the intensity of the trip generation in each zone and its importance to the overall transportation study. These zones generally contain many minor streets and residences that might be considered as nodes or links in a detailed representation of the real network. Each zone, however, is represented by a "centroid," which, in effect, means a single node. This simplification corresponds to "compressing" part of the real graph. It is implied that all nodes within a zone are accessible from each other in the real network.

The approximating network also includes a subset of the complete transportation network consisting of all principal transportation facilities (main roads, transit lines). Junctions or intersections in this network are also represented as nodes, and connecting transportation facilities are represented as links in the approximating network. This subgraph is, presumably, completely connected.

All trips are considered to have origins and destinations at the centroids, which, on a map, seem to be located at artificial points, not on the main roads. The centroids are connected to the main road network by "dummy links" (which are not intended to represent any particular physical facilities). Generally the approximating network does not provide direct links between centroids; however, all routes between centroids must go by way of dummy links to and from the main road network. Altogether the approximating network could be described as a subgraph of a compressed graph of the complete network. The "real" nodes are first partitioned into sets, some of which are centroids and others main road junctions, to form a compressed graph. One may leave out certain links of this compressed graph, however (as defined in section 2.8), particularly those between centroids and possibly some of the links between centroids and certain main road junctions.

As a further complication, it is often postulated that a route between two nodes should not pass through a centroid; that is, centroids can be only end points of routes. If the dummy links can be considered as proxies for some minor roads, this restriction seems unnecessary. On the other hand, the dummy links might sometimes create a nonexistent shortcut in the real network. One obvious way to exclude routes through centroids without modifying the graph-theoretical definitions would be to represent each centroid by two nodes, one of which would be an "origin centroid," the other a "destination centroid." The former would have only one-way dummy links *to* the main roads, the latter would have only one-way dummy links *from* the main roads. In the terminology of section 2.6, the destination centroids would be trap nodes; the origin centroids would be trap nodes for a graph with all one-way links reversed.

Problems

1
For a set of n nodes, evaluate the number of
a. links in a complete directed (undirected) graph;
b. different directed (undirected) graphs that can be constructed;
c. different bipartite graphs;
d. different cut-sets in a complete graph;
e. different routes on a complete graph.

2
For the following graph, construct the accessibility graph. Rearrange both in block triangular form and draw a schematic picture of the graph.

	1	2	3	4	5	6	7	8	9	10	11	12	13	14	15	16	17	18	19	20	21	22	23	24	25	26
1	0	0	0	0	0	0	0	1	0	0	0	0	0	0	0	1	0	0	0	0	0	0	0	0	0	0
2	0	0	0	0	0	0	0	0	0	0	0	0	0	0	0	0	0	0	1	0	0	0	0	0	0	0
3	0	0	0	0	0	0	0	0	0	0	0	0	0	0	0	0	1	0	0	1	0	0	0	0	0	0
4	0	0	0	0	0	0	0	0	0	0	0	1	0	0	0	0	0	0	1	0	0	0	0	0	0	0
5	0	0	0	0	0	0	0	0	0	1	0	0	0	0	1	0	0	0	0	0	0	0	0	0	0	0
6	0	0	0	0	0	0	0	0	0	0	0	1	0	0	0	0	0	0	0	0	0	0	0	0	0	0
7	0	0	0	0	0	0	0	0	0	0	1	0	0	0	0	0	0	0	1	0	0	0	0	0	0	0
8	0	0	0	0	0	0	0	0	0	0	0	0	1	0	1	0	0	0	0	0	0	0	0	0	0	0
9	0	0	0	0	0	0	0	0	0	0	0	0	0	0	0	0	0	0	0	0	0	1	0	0	0	0
10	0	0	0	0	1	0	0	0	0	0	0	0	0	0	0	0	0	0	0	0	0	0	0	0	0	0
11	0	0	0	0	0	1	0	0	0	0	0	0	0	0	0	0	0	0	0	0	0	0	0	0	0	0
12	0	0	0	0	0	0	0	0	0	0	0	0	1	0	0	0	0	0	0	0	0	0	0	0	0	0
13	0	0	0	0	0	0	1	0	0	0	0	0	0	0	0	0	0	0	1	0	0	0	0	0	0	0
14	0	0	0	0	0	0	0	0	0	0	0	0	0	0	0	0	0	1	0	0	1	0	0	0	0	0
15	0	0	0	0	0	0	0	0	1	0	0	0	0	0	0	0	0	0	0	0	0	0	0	0	0	0
16	1	0	0	0	0	0	0	1	1	0	0	0	0	0	0	0	0	0	0	0	0	0	0	0	0	0
17	0	0	1	0	0	0	1	0	0	0	0	0	0	0	0	0	0	0	0	0	0	0	0	0	0	0
18	0	0	0	0	1	0	0	0	0	0	0	0	1	0	0	1	0	0	0	0	0	0	0	0	0	0
19	0	0	0	0	0	1	0	0	0	0	0	0	0	0	0	0	0	0	0	0	0	1	1	1	1	0
20	0	0	0	0	0	0	1	0	0	0	0	0	0	0	0	0	0	0	0	0	0	0	0	0	0	0
21	0	0	0	0	0	0	0	0	0	0	0	1	0	0	1	0	0	1	0	0	0	0	0	0	0	0
22	0	1	0	1	0	0	0	0	0	0	0	0	0	0	0	0	0	0	0	0	0	0	0	0	0	0
23	0	0	0	0	0	0	0	0	0	0	0	0	0	0	0	0	0	0	0	0	0	0	0	0	0	1
24	0	0	0	0	0	0	0	0	0	0	0	0	0	0	0	0	0	0	0	0	0	0	0	0	1	0
25	0	0	0	0	0	0	0	0	0	0	0	0	0	0	0	0	0	0	0	0	0	0	1	0	0	0
26	0	0	0	0	0	0	0	0	0	0	0	0	0	0	0	0	0	0	0	0	0	0	0	1	0	0

3
Suppose that a graph $G_1 = \{N; L\}$ is completely connected. We first remove from L a cut-set $\{X, \bar{X}\}$ and obtain a subgraph

$G_2 = \{N; L \cap \overline{\{X, \bar{X}\}}\}$.
We next remove the cut-set $\{\bar{X}, X\}$ and obtain
$G_3 = \{N; L \cap \overline{\{X, \bar{X}\}} \cap \overline{\{\bar{X}, X\}}\}$.

Determine the form of the graphs G_{2a}^* and G_{3a}^*.

4

What is the least number of links one can have in a completely connected graph on n nodes?

5

Each stop on a bus route is represented as a node. Links join successive stops along the route. Construct the matrix G for the graph representing a bus route that stops at nodes
a. $\ldots, 1, 2, \ldots, n - 1, n, n - 1, \ldots, 2, 1, 2, \ldots$;
b. $\ldots, 1, 2, \ldots, n, 1, 2, \ldots$.
Passengers must pay a fare when they board the bus and when they pass node n. Construct the accessibility graph of nodes that can be reached by a single fare.

6

Suppose that during the football season no team plays another more than twice. At the end of the season, we list all the teams $1, 2, \ldots, n$ and construct a graph G. A link (i, j) in G means that i played against j and j won. If i and j played twice with the same result, G has one link (i, j); if each team won one game, G contains links (i, j) and (j, i). There are no tie games. How would you interpret the graph G_a^* and the block triangular decomposition of the graph?

References   The following books describe the properties of graphs and flows on graphs relevant to transportation network analysis:

1
Ford, L. R., and Fulkerson, D. R. *Flows in Networks.* Princeton University Press, 1962.

2
Berge, C., and Ghouila-Houri, A. *Programming, Games, Transportation Networks.* Methuen, Inc., 1965. Translated from 1962 French edition.

3
Busacker, R. G., and Saaty, T. L. *Finite Graphs and Networks.* New York: McGraw-Hill Book Co., 1965.

4
Potts, R. B., and Oliver, R. M. *Flows in Transportation Networks.* New York: Academic Press, 1972.

Much of the representation and decomposition methods described in this chapter are extracted from similar procedures used in the classification of Markov chains in probability theory.

5
Feller, W. *An Introduction to Probability Theory and Its Applications.* Vol. 1, 2nd ed. New York: John Wiley & Sons, Inc., 1957.

6
Gantmacher, F. R. *The Theory of Matrices.* New York: Chelsea Pub., 1959. Translated from Russian.

3 METRICS AND SHORTEST ROUTES

Two basic concepts associated with transportation networks are those of "distance" between nodes or on links, and flows between nodes or on links. These two notions are quite separate; a network can have a distance but no flow, or flows with no distance. In transportation, however, the two concepts are coupled by the tendency of people to travel by the shortest routes, thus generating flows dependent on distances.

We are not constrained to think of "distance" in the usual physical sense, and, in most applications, we would prefer *not* to do so. In most transportation planning procedures, one usually thinks in terms of travel costs or time, or some linear combination of these, rather than distance.

Abstract distances, or metrics, are familiar notions in mathematics (calculus, functional analysis, topology) and can be defined on quite general types of spaces. Let S be any space with points $\alpha, \beta, \gamma, \ldots$. A *distance*, or *metric*, on S is a real-valued function d defined on pairs of points in S with the following properties for all $\alpha, \beta, \gamma \in S$:

positive definite $\qquad d(\alpha, \beta) \geq 0, \quad d(\alpha, \beta) = 0$ if and only if $\alpha = \beta$;

symmetric $\qquad d(\alpha, \beta) = d(\beta, \alpha)$; and

triangle inequality $\quad d(\alpha, \beta) + d(\beta, \gamma) \geq d(\alpha, \gamma)$.

The usual motivation for introducing metrics in rather general spaces is to define the notion of *limits*. Given that one has already defined limits for a sequence of real numbers, one can define

$$\alpha_j \to \alpha \text{ for } j \to \infty$$

to mean

$$d(\alpha_j, \alpha) \to 0,$$

where $d(\alpha_j, \alpha)$ is a sequence of real (nonnegative) numbers. Without a metric we can only say that either $\alpha = \beta$ or $\alpha \neq \beta$; we cannot say that α is "close" to β or that a sequence of points α_j "approaches" β. For example, in functional analysis an element of S may be a real-valued function of a real variable $g(x)$, designated here simply by a label α. For various definitions of d, one can define the convergence of a sequence of functions, $g_j(x) \to g(x)$, in various ways.

Each branch of mathematics has its own favorite examples of metrics peculiar to the problems encountered in its area. In many branches of mathematics a metric is perhaps an artificial notion in terms of which one can formalize different definitions of limits; in transportation or on networks metrics are usually an indication of the amount of effort (in some sense) required to go from one place to another.

In network analysis the metrics considered are, on the one hand, more specialized than in other fields in that they have special structure (for example, the distance along a route is usually the sum of the distances on its parts); on the other hand, they are more general in that the symmetry axiom may be abandoned. The distance from α to β along directed links might not be the same as from β to α. In the remainder of this section we shall explore several examples of metrics.

1. The most primitive function that satisfies the axioms of a metric is

$$d(\alpha, \beta) = \begin{cases} 1 & \text{if } \alpha \neq \beta, \\ 0 & \text{if } \alpha = \beta. \end{cases} \tag{3.1}$$

Even this simple function has some relevance to transportation as a component of trip cost, namely the cost of initiating the trip. It might cost one unit to make a trip, nothing if one does not.

2. In an undirected graph, with nodes $i, j, \ldots$

$$d(i, j) = \begin{cases} \text{minimum number of links on any route from} \\ i \text{ to } j, \\ \infty \text{ if } j \text{ is not accessible from } i. \end{cases} \tag{3.2}$$

On a directed graph, we could use the same definition, but it would not necessarily be true that $d(i, j) = d(j, i)$. Note that on some idealized networks (such as a square grid) with unit distance on all links, this might also represent the travel distance or travel time. For a complete graph this $d(i, j)$ is the same as example 1.

3. On a square grid of roads with block length a, let nodes have Cartesian coordinates $[ak, al]$, where k and l are integers.* In defining $d(\alpha, \beta)$, the points $\alpha, \beta, \ldots$ are represented as ordered pairs of real numbers ak and al. Suppose we define

$$d([ak, al], [ak', al']) = |ak - ak'| + |al - al'|. \tag{3.3}$$

The definition of a metric in transportation is usually chosen so that the distance is identified with the cost of travel along some "best" route. The abstract definition of a metric says nothing about routes. The metric is associated with pairs of nodes only; d is a function of pairs α, β. There is no mention of the existence or nonexistence of links. In this example, however, the metric is clearly identified with minimum journey distance, although the route is not unique. All routes with monotone varying coordinates along the route have the same travel distance.

4. Example 3 is a special case of a metric that is used in many branches of mathematics. Suppose S is a linear vector space, the elements of which are ordered sets of real numbers $x_j (j = 1, 2, \ldots, n)$. If the point α is represented by the ordered sequence

$[x_1, x_2, \ldots, x_n]$, and β by $[y_1, y_2, \ldots, y_n]$,

*The symbol (x, y) is commonly used to represent an ordered pair of objects x and y, with x in one space and y in the same or a different space. Thus in the representation of points in the plane, the symbol (x, y) represents the Cartesian coordinates. Unfortunately, we have already used the symbol (α, β) to represent the links between α and β, which is also an ordered pair of objects. Because we will be using the symbol (α, β) to represent the link when α is a point in a two-dimensional space, we shall use the symbol $[x, y]$ to represent the points in the plane.

then we can define

$$d(\alpha, \beta) = \sum_{j=1}^{n} |x_j - y_j| \tag{3.4}$$

or, more generally yet,

$$d(\alpha, \beta) = \left\{ \sum_{j=1}^{n} |x_j - y_j|^\gamma \right\}^{1/\gamma} \tag{3.5}$$

for some $\gamma \geq 1$. The case $\gamma = 1$ corresponds to (3.4); the case $\gamma = 2$ and $n = 2$ or 3 is the usual Euclidean distance in two or three dimensions. For $\gamma = 1$, $n = 2$, this metric differs from (3.3) only in that the space in example 3 consists only of points on a square lattice. If, however, the length a is arbitrarily small, there is a lattice point arbitrarily close to every point in the plane, so that the present metric is essentially equivalent to (3.3).

Even though it is possible to follow a route from α to β that stays arbitrarily close to a straight-line path in the Euclidean sense, the distance of travel (3.4) is not the Euclidean length of the straight line. Generally, the travel distance between points that are far apart compared with the spacing of roads depends on the "local geometry" of the network.

5. Most metrics of interest in transportation have a special mathematical form suggested by example 4. They are defined on graphs that contain not only nodes but also links. Instead of explicitly listing $d(i, j)$ for all $i, j \in N$, we specify some distances on links and a scheme for computing a metric $d(i, j)$. Let us define a function $d_{ij} > 0$ for all $(i, j) \in L$; we can interpret d_{ij} as the "length" of the link (i, j) or the distance from i to j via the link (i, j).

For any route R composed of links (n_1, n_2), (n_2, n_3), ..., (n_{r-1}, n_r), we define the length of R by

$$d(R) = d_{n_1 n_2} + d_{n_2 n_3} + \ldots + d_{n_{r-1} n_r}; \tag{3.6}$$

that is, the length of a route R is the sum of the lengths of its component links. We now define as our metric

$$d(i, j) = \min_{R \mid n_1 = i, \, n_r = j} d(R); \tag{3.7}$$

that is, it is the length of the shortest route from i to j. For $i = j$, we define $d(i, i) = 0$; and if there is no route from i to j, we define $d(i, j) = \infty$.

Except possibly for the symmetry property, this $d(i, j)$ satisfies all the axioms of a metric. For $i \neq j$, it is a sum of positive terms, so that $d(i, j) > 0$. To prove the triangle inequality, let i, j, and k be any three nodes. Let R_1 be a route from i to j such that $d(i, j) = d(R_1)$; that is, R_1 is a shortest route from i to j. Also let R_2 be a shortest route from j to k. The route $R_1 R_2$ is a route from i to k and, furthermore, by definition (3.6),

$$d(R_1 R_2) = d(R_1) + d(R_2) = d(i, j) + d(j, k).$$

But $d(i, k)$ is the minimum length of all routes from i to k and is, therefore, at least as small as any particular route; thus,

$$d(i, k) \leq d(R_1 R_2) = d(i, j) + d(j, k).$$

One of the key features of this last metric is the additivity of length on links. Although one might define the d_{ij} to include factors such as time, money, or inconvenience, it is not obvious that any such additive metric is a realistic representation of the cost or difficulty of traversing a route. Neither is additivity a necessary condition for a function $d(i, j)$ to satisfy the axioms of a metric.

There are two obvious weaknesses of the additivity on links. First, there may be costs associated with factors such as turns or transfers between modes at the nodes. Second, short trips are typically more expensive (or take more time) per unit distance than long ones. The first weakness is discussed in the literature. It is usually resolved by the introduction of extra nodes and links or, otherwise, by the addition of some internal structure to the nodes. However, because a $d(i, j)$ need not be identified with a route (it is, in fact, just a function on nodes), it is not necessary to complicate a graph merely for the purpose of defining a metric. Most schemes for including costs at nodes merely add to the above metric a few additional terms, preserving the additivity of costs along routes.

The second weakness of additivity may be more basic.

Although its existence is generally recognized, to my knowledge no one has attempted to explore its consequences to transportation planning. Money costs, such as for fuel or maintenance, probably are approximately additive but could perhaps be made somewhat more realistic by addition of an extra cost for initiating a trip (a cost independent of length except for length 0), as in example 1 above. Furthermore, if $d_1(\alpha, \beta)$ and $d_2(\alpha, \beta)$ are any two metrics, then

$$d(\alpha, \beta) = a d_1(\alpha, \beta) + b d_2(\alpha, \beta) \tag{3.8}$$

is also a possible metric for real positive numbers a and b. In particular, any linear combination of the metrics 1 and 5 defines a metric.

If $d_1(\alpha, \beta)$ represents a metric associated with money costs, and $d_2(\alpha, \beta)$ a metric associated with time (both of which are additive along the route with possibly an extra penalty to initiate a trip), new metrics can be generated by linear combinations of these; the ratio b/a can then be interpreted as the "cost equivalent of time." Although this is mathematically acceptable, it is not necessarily realistic. The difficulty is that any sensible cost equivalent of time should depend on trip length, thus destroying the additivity of the metric. One minute of time added to an hour-long trip does not have the same "value" as a minute added to a one-minute trip. An equivalent but perhaps more convincing interpretation is that ten people saving one minute each, or one person saving one minute on each of ten trips, is not the same as one person saving ten minutes on a single trip.

Realistic measures of trip cost might still satisfy the axioms of a metric because even along a single route one would probably now postulate that

$$d(R) \le d_{n_1 n_2} + \ldots + d_{n_{r-1} n_r},$$

which would continue to satisfy a triangle inequality.

**3.2
Shortest Path,
Huygens's Principle**

The most commonly used metric in transportation has the form described in example 5 of section 3.1. On any graph having a finite number of nodes, there are a finite number of routes between nodes i and j; consequently, it is, in principle, a simple matter to find the route with the shortest distance. One can simply evaluate the lengths of all routes and choose the best. In practice, however, this method is not only inefficient but usually impossible. On a complete graph of n nodes, the number of routes from i to j is on the order of $(n - 2)!$ (if we are wrong by a factor of 3 or 10, it doesn't make much difference). For $n \approx 20$ this is already far too many routes for any present-day computer to handle. But the situation is not quite this bad. In most transportation networks, each node will have only a few links. We exclude from consideration paths that go from i to j to i (or any larger cycles); thus there are only perhaps two or three possible "next links" in a route. Thus there might be about c^n routes from i to j, where c is a number on the order of 2. A computer could possibly examine all routes on a graph of 30 nodes.

There are, in fact, very efficient algorithms for the evaluation of shortest routes in a graph, even for networks with $n = 10^3$ or 10^4. In network analysis it is quite common that a calculation one tries to perform with "zero" thought will require an impossible amount of computation. Some thought will reduce the computation by a factor of maybe 10^{10} or so, and maybe a little more thought will make the calculations trivial (they can be done by hand or with a geometric construction).

The particular problem of finding the shortest route between two points in a space is one of the oldest and most thoroughly analyzed in all of mathematics. Unfortunately, many of the very elegant ways of looking at this problem have been lost because of the recent fashion of programming all computations for computers.

The original shortest-path problem must have been associated with the notion that the shortest path between two points in physical space is a straight line. This led to questions of shortest paths on a sphere, on general surfaces in three-dimensional space, and finally to the branch of mathematics known as differential geometry. In physics, almost every "basic" law has a formulation in terms of a minimum principle. In geometric optics, a light ray follows a path of minimum travel time (even if the velocity of light or index of refraction changes from place to place). A dynamical system follows a trajectory that minimizes the time integral of a Lagrangian. Optimal-path problems can be interpreted to be the motivation for much of the vector calculus and the calculus of variations.

Most shortest-path problems in classical applied mathematics deal with paths in a two- or three-dimensional continuum. As an example, a two-dimensional version of the analysis of light rays in geometric optics might be formulated as follows. At every point $[x_1, x_2]$ in the two-dimensional space there is a local velocity of light $v(x_1, x_2)$, which depends on the properties of the medium. There is also a local metric such that the distance (actually the travel time) from $[x_1, x_2]$ to $[x_1 + \varepsilon_1, x_2 + \varepsilon_2]$ is approximately

$$\frac{(\varepsilon_1^2 + \varepsilon_2^2)^{1/2}}{v(x_1, x_2)}$$

for sufficiently small $[\varepsilon_1, \varepsilon_2]$; that is, the Euclidean distance divided by the local velocity. The length of any path is the sum of the distances along sufficiently close points along the path, which is the integral of $[v(x_1, x_2)]^{-1}$ along the path. The distance from $[x_1, x_2]$ to $[y_1, y_2]$ is the distance along the shortest path, and a "light ray" is defined to be this path.

The only difference between this metric and those described in section 3.1 is that the space here forms a continuum of an infinite number of points, and the distance on links is replaced by a distance between neighboring points in the continuum.

The standard method of constructing rays in geometrical optics is to start from an origin (light source) and draw a circle (sphere in three dimensions) of "small" radiums ε, chosen so that the velocity $v(x_1, x_2)$ can be considered as essentially constant over the circle $x_1^2 + x_2^2 \leq \varepsilon^2$. This circle is the (approximate) locus of all points that can be reached in time $\varepsilon/v(0, 0)$; that is, the points are a "distance" $\varepsilon/v(0, 0)$ from the source. A light ray from the origin to a point on the circle is (approximately) the straight-line path.

Subsequent steps are performed by an iteration procedure known as Huygens's principle, named for the seventeenth-century Dutch physicist and mathematician Christian Huygens. Suppose that at the ith stage we construct a curve T_i of all points that can be reached in time T_i from the origin; for example, the above circle for $i = 1$. (Such a curve is called a wave front in optics literature.) Suppose we also find all shortest paths (rays) to points on T_i. At each point $[x_1, x_2]$ on T_i, there is a local velocity $v(x_1, x_2)$. In a time $T_{i+1} = T_i + \delta$, it is possible to reach any point on a circle of radius $v(x_1, x_2)\,\delta$ around the point $[x_1, x_2]$ by traveling along a ray to $[x_1, x_2]$ and a straight line from $[x_1, x_2]$ to the circle. Geometrically it is clear that if T_i is a smooth curve, the collection of all circles centered at points $[x_1, x_2]$ on T_i will have an envelope forming another curve T_{i+1} that encloses T_i, as shown in figure 3.1. All points on the envelope can be reached in time T_{i+1}, and all points between T_{i+1} and T_i can be reached in a time less than T_{i+1}.

To construct a ray from the origin to a point $[x_1', x_2']$ on T_{i+1}, we choose the circle in the geometric construction that is tangent to T_{i+1} at $[x_1', x_2']$ and find its center. Now we extend the ray from 0 to the circle center on T_i, with a straight line to T_{i+1}. Having constructed T_{i+1} and rays to T_{i+1} from T_i we can proceed to find T_{i+2}, T_{i+3}, and so forth.

This procedure is of more than historical interest; "modern" methods of doing shortest-path calculations on a transportation network may not be identical to this, but most are very similar to it. They usually start from the origin and systemically identify points that are farther and farther away. In the process, one obtains not just the distance and path to a single point but the distances and paths to all points (which one may eventually need anyway). One also uses (indirectly, at least) what, in the modern terminology, would be called a principle of dynamic programming. In the present context it would say that if a shortest path from α to γ passes through point β, then the path α-β-γ must contain a shortest path from α to β and from β to γ. In effect, we applied this principle when we identified a ray to T_{i+1} as the union of a ray to T_i and a ray in a circle (shortest path) between T_i and T_{i+1}.

Sometimes one can apply a dynamic programming argument in a more direct way. In deriving a law of refrac-

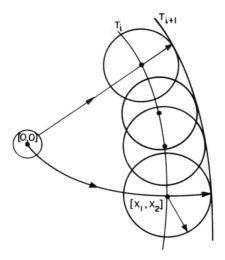

Figure 3.1
Huygens's construction of rays
and wave fronts.

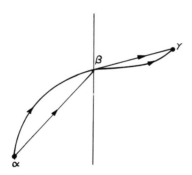

Figure 3.2
Snell's law of refraction.

tion of light, we imagine that we have two types of media
with a boundary that is at least locally straight. In one
medium the velocity is v_1, in the other v_2. We have a source
(origin) in one medium at point α, and we wish to find the
shortest path (ray) to a point γ in the second medium. A
procedure for doing this might be to consider an arbitrary
path α-β-γ, as shown in figure 3.2, and then systematically
seek shorter paths until there is no possible improvement.

The first step might be to observe that if the path from α to β in the first medium has any curved portion, then there is a shortcut. We are quickly led to the conclusion that the path from α to β, with β on the boundary, must be a straight line. (This follows also from the Huygens construction.) The path from β to γ must also be a straight line, and the shortest path from α to γ must pass through some point β. The final determination of β is a simple exercise of minimizing a function (the distance along α-β-γ) of the single parameter β. The solution is called Snell's law of refraction (also formulated in the seventeenth century).

**3.3
Shortest-Path
Algorithms**

Most great ideas in mathematics originate from simple geometric or schematic pictures. These pictures are then translated into a formal logic of axioms, theorems, and so forth. These, in turn, are generalized to situations in which the pictures may not be apparent anymore. Thus the fields of differential geometry, calculus of variations, and dynamic programming have evolved from problems like those described earlier.

Computer algorithms frequently evolve in the same way, except that the "language" is different not only in the detailed sense of how the program is written but also in the structure. In "abstract" mathematics, multidimensional spaces are no problem, and there is no necessity of casting the logic in an "ordered" space. A computer, however, is even less flexible than a geometric picture. A computer program is a translation into a one-dimensional logic or, actually, an integer sequential logic. The computer works basically on the principle, "go from step i to $i + 1$," repeated a finite number of times (preferably not exceeding 10^{10}).

What one does in mathematics is, in a certain sense, almost opposite to what one does in programming. In mathematics one seeks a sophisticated, compact notation for general concepts. In programming one seeks a very primitive but systematic procedure with as few sequential steps as possible.

A very large number of computer programs have been written for finding shortest routes. Most of these are based on starting at the origin and systematically finding distances and shorter paths to points farther and farther away (Huygens's principle); or proposing tentative routes and systematically improving them; or a combination of these two procedures.

In translating geometric constructions into computer language, there are two rather trivial steps with which the computer has trouble. First, a computer does not instantly recognize what points are near others. It can do this in one dimension by ordering points, but in two or three dimensions there is no simple way of ordering nodes so that the computer immediately recognizes that a node can have close neighbors in several directions. It must systematically go through all nodes and inspect the table of distances.

Second, a computer can handle only a finite number of nodes. Although the Huygens construction is a finite geometric algorithm for determining approximate distances, it exploits some simple features of a two-dimensional continuum that the computer has difficulty recognizing. In particular, the construction exploits the fact that the curve of constant travel time T_i is a connected curve. It divides the space into an inside and an outside such that every path from the inside to the outside must cross T_i. Furthermore, every point in the region between the curves T_i and T_{i+1} has a travel time between the values T_i and T_{i+1}. Thus, for sufficiently small $T_{i+1} - T_i$, one can approximate the travel time as accurately as one pleases for the infinite number of points between the curves. For any finite but large subset of points between the curves, a computer algorithm would need to inspect the points one at a time.

The following two examples of shortest-path algorithms, rather than giving a survey of the most refined methods (which is, indeed, a subject in itself), illustrate the type of reasoning used in the development of efficient computation schemes.

1. This algorithm was historically one of the first. It is not, by present standards, very efficient, but it most clearly mimics the type of logic used in the Huygens construction.

We wish to find the shortest path from one node, which we arbitrarily designate node 1, to all other nodes. First, the computer is given a listing of all nodes i, links (i, j), and distances on links d_{ij}. Some memory locations are set aside in which we can eventually record whether node i has been "labeled" (yes or no); the distance d_i of node i from the origin; and the node n_i from which node i is reached by a link (n_i, i) on the shortest path (or several nodes if the shortest path is not unique).

Any efficient computer algorithm will give computation instructions in such a form that the same instructions are used over and over again. A computer can follow the instruction "go from step k to $k + 1$" in 10^{-7} second or less. It can read punched cards at a much slower rate, but writing the instructions and punching the cards must, typically, be done by hand. If this cannot be done quickly (because there are too many instructions), the speed of the computer is largely wasted.

The present algorithm proceeds as follows. For step $k = 1$, label node 1 and set $d_1 = 0$, $n_1 = 1$. (The designation $n_1 = 1$ is redundant but serves further to identify node 1 as the origin.) For steps $k = 2, 3, \ldots, n$, let X_{k-1} be the set of labeled nodes at step $k - 1$ and $\bar{X}_{k-1}$ be its complement. Calculate

$$\min_{i \in X_{k-1}} \left[d_i + \min_{(i, j) \in (X_{k-1}, \bar{X}_{k-1})} d_{ij} \right] \tag{3.9}$$

and observe the particular values of i and j for which the minimum is achieved. Now change X_{k-1} to X_k by adding to X_{k-1} the node j that gave the minimum. Node j is assigned the distance from (3.9) as d_j and node i as n_j. If there is more than one node i that yields the same value of d_j, one may either choose one of them or list them all. Increase k to $k + 1$ and repeat until all nodes are labeled.

In more conventional language, the set X_{k-1} of labeled nodes is the set for which the shortest path from 1 has already been evaluated. Furthermore, it contains the $k - 1$ nodes nearest the origin (including the origin node itself). Each node n_i is a node from which i was reached on a shortest path. The specification of a node n_i for each node in X_{k-1} enables us to trace the path to i back one link to a previous node, from which it can be successively traced back one step at a time to node 1. In effect, the set X_{k-1} and the nodes n_i, $i \in X_{k-1}$, are the analogs of the interiors of T_i and its rays in the Huygens construction. This iterative procedure is one by which, having found the $k - 1$ nearest nodes to node 1, one finds the kth nearest.

Because it is assumed that $d_{ij} > 0$ for all links, any path from 1 to j must pass through nodes nearer to 1 than j (or be a direct link). In particular, the kth nearest node to node 1 must have a link from X_{k-1}. In this procedure, $d_i + d_{ij}$, where $i \in X_{k-1}, j \in \bar{X}_{k-1}$, and $(i, j) \in (X_{k-1}, \bar{X}_{k-1})$, is a distance to node j via a shortest route to i and a direct link from i to j (i may be 1, and $d_1 = 0$). The square bracket in (3.9) identifies the nearest unlabeled node to i and its distance from node 1 via i. The minimum over i identifies the nearest unlabeled node to 1 via any path through X_{k-1}, which, in turn, is the nearest of all unlabeled nodes.

This procedure is computationally tedious because the computer does not know the best place to look for the kth node. It must search all links (i, j) in the cut-set $(X_{k-1}, \bar{X}_{k-1})$ and compare the distances to find the smallest one. This difficulty cannot be avoided. This computer algorithm is not the best, however, because in the $(k + 1)$th step one may need to compare many of the same lengths (plus some new ones) that one used in the kth step. The second-best distance in the kth step might be the best at the $(k + 1)$th step. The implication in the algorithm is that the computer discards all such intermediate calculations not relevant to the immediate objective (namely to find the kth nearest node) and must recalculate them if they are needed later.

It is, of course, typical of any hand or computer calculation that to keep intermediate results, one must not only have a place to put them but have some kind of filing system so that they can be found again. For a computer, one must write a longer program of instructions.

If one does not attempt to save intermediate results, expression (3.9) can be evaluated in the following way. First, if each node i has many links $(i, j) \in L$, it may be advantageous to list them, for each i, in order of increasing d_{ij}, so that the second term of (3.9) can be evaluated simply by selecting the first link (i, j) that is also in $(X_{k-1}, \bar{X}_{k-1})$.

(Usually there are not very many links from i to any j, so the evaluation of the second term is not much of an issue.)

To find the minimum with respect to i in (3.9), the standard procedure is to set aside a memory location in which one records a number d, initially set very large. At step i one evaluates $d_i + \min d_{ij}$ for the ith node of X_{k-1}. If this new number is smaller than the existing value of d, it is inserted in place of that value. Now i is changed to $i + 1$, and the process is repeated until $i = k - 1$. The final value of d is the value of (3.9).

Before doing any calculations, it is important to make preliminary estimates of computation time to see whether the cost of the calculation is trivial or larger than the national debt. The time required to perform elementary operations, such as comparing two numbers, extracting numbers from the (fast) computer memory, changing indices in an iteration, or reading an instruction, is relatively small (about 10^{-7} second) compared with the time required to perform an addition (10^{-6} second), which, in turn, is smaller than the time required to do a multiplication or a division (10^{-5} second). Usually in estimating computation time, one neglects the time required for elementary operations and additions and only estimates the number of multiplications or divisions (or the number of times one must use various subroutines of known longer computation time) and multiplies by the time per evaluation.

In these shortest-path algorithms there are no multiplications. The slowest operation is an addition, but if one is not careful, one may find the total computation time dominated by a much larger number of comparisons than additions.

To evaluate (3.9) one must perform one addition for each i in X_{k-1} and do a certain number of comparisons to select the second term. If the latter can be neglected, the total time can be estimated from the fact that, for each step k, one must perform $k - 1$ additions. The total number of additions in all steps is therefore approximately

$$\sum_{k=2}^{n} (k - 1) \approx \frac{n^2}{2}.$$

For most transportation networks, however, the graph has a two-dimensional geometry similar to that in the Huygens's construction. In particular, each node has links with only three or four other nodes, which are neighbors in a two-dimensional Euclidean sense; the $k-1$ nodes nearest the origin, X_{k-1}, are the nodes within some area A_{k-1}.

In the evaluation of (3.9) the only nodes $i \in X_{k-1}$ that have links $(i,j) \in (X_{k-1}, \bar{X}_{k-1})$ are nodes near the "surface" of A_{k-1}. As one goes through the algorithm, X_k always includes X_{k-1}. If at any step k one finds that a node i has no links to $\bar{X}_{k-1}$, it can be eliminated from subsequent evaluations. It will never again enter into an evaluation of (3.9) for larger k. Also, if there is any link $(i,j) \in L$ with both i and j in X_{k-1}, it can be eliminated from subsequent evaluations.

If nodes are distributed more or less uniformly over a two-dimensional plane, the length of the surface of A_k will be proportional to $k^{1/2}$. The number of links in $(X_{k-1}, \bar{X}_{k-1})$ will also be proportional to $k^{1/2}$, perhaps about $2\,k^{1/2}$. A more realistic estimate of the number of additions in the algorithm would be

$$\sum_{k=2}^{n} 2k^{1/2} \approx \frac{4n^{3/2}}{3} \approx n^{3/2}.$$

If an addition takes 10^{-6} second of computer time and $n = 10^4$, the computation time (for a single origin node) is about $10^{-6} \times (10^4)^{3/2} = 1$ second. If the number of additions were $n^2/2$, the computation time would be about $10^{-6} \times (10^4)^2 = 100$ seconds.

2. In the following algorithm we keep intermediate calculations that may be useful in later stages of the calculations and simplify the program as well.

In this procedure there are again two classes of nodes, but, instead of being unlabeled or labeled, they have either temporary or permanent labels. Let X_{k-1} be the set of nodes with permanent labels at step k. Initially each node j is given a temporary label, a temporary distance d_{1j}, and a previous node label $n_j = 1$. If $(1,j)$ is not in L, set d_{1j} equal to some very large number (∞). Make node 1 the set X_1 with permanent labels, and give it the distance $d_1 = 0$.

For $k = 2, 3, \ldots, n$, choose node $j \in \bar{X}_{k-1}$ with the smallest temporary distance. Add node j to X_{k-1} to form X_k; for this single node j, and for each $i \in \bar{X}_k$ and $(j, i) \in L$, calculate $d_j + d_{ji}$. If this number is less than the existing temporary distance for node i, make this the new temporary distance, and make j the new previous node label n_i for node i. Change k to $k + 1$ and repeat until all nodes have permanent labels and distances d_i.

This procedure generates exactly the same sets X_k as the algorithm in (1), with the same interpretations as the k nearest nodes to 1. The calculations are simpler, however, because there is a much smaller number of distances to computer at each stage k. Having chosen the single node j, one need only consider the nodes i with direct links from j (of which there are probably only three or four) and, of these, only the nodes with $i \in \bar{X}_k$. The simplification is achieved because at each step one systematically keeps in the table of temporary distances any computed values of $d_j + d_{ji}$ that might be useful (as permanent distances) at a later stage, and discards any previously computed values that are no longer of any value. The critical point is that there is a simple scheme to store, classify, retrieve, and discard intermediate calculations that may or may not be useful later; it is sufficiently simple that it is faster to retrieve previous results than to recalculate them.

To prove that the permanent labels are the shortest distances, one need only observe that, after step k, each node j in $\bar{X}_{k-1}$ has a temporary distance that is the shortest distance to j by way of any direct link from X_{k-1}. The smallest of these temporary distances must be the shortest distance from node 1 to any node in $\bar{X}_{k-1}$, for the same reasons as in the first algorithm.

The number of additions at stage k is typically only about 2; the total number of additions for all k is, therefore, about $2n$ (less by a factor of $n^{-1/2}$ from the previous algorithm). For $n = 10^4$, the calculation that took one second by the previous algorithm should now take only about 2×10^{-2} second. To achieve this speed, however, one must make sure that the time to retrieve numbers from the memory is kept low compared with the addition time. In particular, the search for the node $j \in \bar{X}_{k-1}$ with the smallest temporary distance must be done without inspection of all distances in $\bar{X}_{k-1}$ (which would require much more time than the additions). The nodes must be stored in the memory (perhaps in order of their temporary distances from the origin) so that one can find this node immediately.

There are other schemes that are usually less efficient than these two algorithms. In essence, these involve assigning some arbitrary routes from node 1 to every j and determining their distances. One then systematically tries to improve the routes by "short cuts" by way of direct links from other routes.

**3.4
All Nodes to All
Nodes**

In analyzing distances on a graph, one usually wants the distances not just from a single node to all other nodes but from every node to every other node. One way to obtain this information is simply to repeat the procedure of section 3.3 with each node i relabeled as node 1. It is obvious, however, that this involves considerable duplication of calculation. In the calculation of the shortest path from 1 to j, one has constructed a route from 1 to l to j, say, such that the routes from 1 to l and from l to j are also shortest routes (the latter is of particular interest). Now, to construct routes from l to all points, one should be able to use the fact that some of the routes have already been determined.

There are computer algorithms that treat all nodes at once without just repeating the algorithms of section 3.3 for each origin node, but they do not seem to yield the speed one might expect (they may be two or three times faster). Unfortunately many tricks that are recognized and exploited instinctively in a hand calculation or geometric construction (for example, the optimal path obviously does not go out in the "wrong" direction) are difficult to program for a computer and may not be universally applicable.

In transportation studies, trips are usually considered to start and end only at centroids, as described in section 2.9. Although all nodes must be considered in the calculation of shortest paths from a single centroid to all centroids, the calculation must be repeated only from the set of centroids, not all nodes. Whereas the total number of nodes may be of the order 10^3 or 10^4, the number of centroids is typically only a few hundred or less. The entire calculation of shortest paths from all centroids can usually be done within a minute or so on a fast computer. To exclude trips passing through centroids, one can consider each centroid as two nodes, an origin centroid and a destination centroid, each having only one-way links to the main network. In calculating shortest paths from a single origin centroid, however, one need not even list the other origin centroids because they are not accessible.

3.5 Other Methods

By far the most tedious parts of a calculation of shortest paths on a transportation network are selecting an approximating network, labeling nodes and links, evaluating distances on links, and punching computer cards to feed these data into the computer. If one did not need to "code" the network for other calculations such as trip distribution or assignment, or compare shortest paths on a wide variety of modifications of the network, but only wanted to know shortest paths and distances between various points on a specified network with a single set of link distances, one might look for cheaper ways to calculate these distances or be satisfied to have some crude estimates from more readily available data.

If one had a street map of a city and wished to determine the shortest-distance (not necessarily the fastest) route between any two points, one could, in fact, usually find it rather quickly. One would look for streets that run obliquely to grid directions, more or less in the direction between the two points, and identify obstacles such as bridges. A few likely candidates for best route can be quickly identified, and a few route distances actually measured to determine the shortest route by direct comparison. To consider trip time rather than distance, one would look for easily accessible freeways or arterials. This is how most travelers, in fact, choose routes, at least in a first trial. They would also see that the differences between the most likely best routes are usually so small as to be inconsequential. People can immediately identify many things on a map and make a variety of judgments that are quite tedious to formulate for a computer.

Suppose one wanted to find the shortest distance between many pairs of nodes from a street map with undirected streets and $d_{ij} = d_{ji}$. One could lay some flexible string along every street and tie a knot at each intersection, like a fish net. If one were to grab the ith and jth nodes and pull the net until the strings were taut, the constraint that would determine how far apart the nodes could be pulled would be the shortest length of string between i and j. For an "arbitrary graph" of many nodes and links, but $d_{ij} = d_{ji}$, one can, in principle, generalize this by cutting the strings so that the length of string along a link (i, j) is d_{ij} (not necessarily the physical length of a street). The resulting network of strings could become a useless pile of knots because of the possible multiple layers of string; however, for a real street network, which is nearly a planar graph, it will probably be no worse than a fish net with some uneven threads.

Problems   1

Show that the metrics proposed in paragraphs 1, 2, and 3 of section 3.1 actually satisfy the axioms of a metric. Also show that $ad_1 + bd_2$ is a metric if d_1 and d_2 are metrics and a and b are any positive numbers.

2

Let $g(x)$ be any real-valued function of a real variable x satisfying the conditions

$$0 = g(0) < g(x) \le g(x + y) \le g(x) + g(y), \qquad x > 0, \quad y \ge 0.$$

a. If $d(\alpha, \beta)$ is a metric on any space S, show that $g(d(\alpha, \beta))$ is also a metric on S.
b. Show, for example, that

$$g(x) = \begin{cases} a + bx, & x > 0, \\ 0, & x = 0, \end{cases}$$

and

$$g(x) = ax^{1/2}$$

satisfy the above conditions for any positive numbers a and b.

References For an introduction to metric spaces and their applications in mathematical analysis, see

1
Kolmogorov, A. N., and Fomin, S. V., *Elements of the Theory of Functions and Functional Analysis*. Vol. 1. Metric and Normed Spaces. Rochester, New York: Graylock Press, 1957. Translated from 1954 Russian edition.

The use of the Huygens construction and other techniques of classical applied mathematics to describe shortest paths is discussed in

2
Wardrop, J. G., "Minimum Cost Paths in Urban Areas," *Strassenbau und Strassenverkehrstechnik*, Heft 87, pp. 184–189. Bonn, 1969. Proceedings of the Fourth International Symposium on the Theory of Traffic Flow, Karlsruhe, 1968.

3
Wardrop, J. G., "Minimum Cost Paths When the Cost per Unit Length Depends on Location and Direction," *Traffic Flow and Transportation*, Proceedings of the Fifth International Symposium on Traffic Flow and Transportation (Berkeley, 1971), pp. 429–438. New York: Academic Press, 1972.

There is a very large amount of literature on shortest path algorithms on discrete networks. Much of this is summarized in previous references on networks, graphs, and so forth, in chapter 2, and in

4
Dreyfus, S. E., "An Appraisal of Some Shortest-Path Algorithms," *Operations Research* 17 (1969): 395–412.

For a discussion of computation times of shortest paths on transportation networks, see

5
Van Vliet, D., "Improved Shortest Path Algorithms for Transport," *Transportation Research* 12 (1978): 7–20.

4 APPROXIMATIONS AND IDEALIZATIONS

4.1
Introduction

Since most real transportation networks contain much more structure than can be handled by any computer, the actual network must be replaced by a skeletal representation. The representation one should use depends on the sorts of questions one would like to answer. Here we will be concerned mainly with the following more or less well-posed mathematical questions. First, given a certain number of trips between various origins and destinations (per unit time) and the cost of travel (distance) on each link, how will the traffic distribute itself over the network if all trips go by the shortest routes? Second, how does this distribution change when various features of the network change?

Present practice is to divide the city or region into zones and represent each zone by a single node (centroid), at least for the purposes of generating traffic. Main intersections or interchanges are also included as nodes, and main roads are included as links. In some cases, one may treat collections of parallel minor roads as a single road of large capacity but slow speed.

Now that we have introduced a metric on graphs and can at least imagine what a suitable metric would be on the real transportation network, we can discuss how a proposed simplified network is an approximation to the real network. For example, we can say that a graph is an approximation to the real transportation network if each origin or destination on the real network can be mapped into an origin or destination on the simplified graph (real origins map into centroids) and the distance between any pair of points on the real graph is nearly equal to that on the image graph. Also, the shortest route on the real graph (defined by the sequence of nodes along the route) should map into a shortest route on the image graph. For the metrics on these graphs, one can invent many possible "global measures" of nearness of one graph to another (for example, the root mean square difference of the distance between corresponding points on the two graphs averaged over all O-D pairs, or the largest difference), but it is better to leave the definition vague, recognizing that certain types of errors are more important relative to certain questions than to others.

If one could divide a study region into sufficiently many zones, one could approximate the real network to any desired degree of accuracy. However, despite their speed, computers cannot handle networks as large as one would like (by present procedures). For various reasons, most large-scale transportation studies can handle only 300–400 zones (in a study region covering perhaps 1,000 square miles). This division yields $(300)^2 \approx 10^5$ origin-destination (O-D) pairs, a number comparable in magnitude to the number of trips analyzed in a base-year O-D survey, but a rather coarse representation of some areas. If zones in subregions of high trip concentration are smaller than one square mile in area, others may be as large as a hundred square miles or more. The number of nodes in the graph, including road intersections, is typically much larger than the number of centroids; it may be in the range 10^3–10^4.

Clearly the accuracy of any approximation of a real network by a skeleton depends on the length and location of the trips being considered. If two graphs are compared in terms of their metrics, two trips between a given pair of centroids must have a travel distance nearly independent of the precise location of the origin and destination within the zone; consequently, the distance between points can be no more accurate than the linear dimensions of the zones. For trips within a zone or between neighboring zones, the accuracy of an approximate graph is very poor; and a significant fraction of trips are, of course, of this type.

Part of the difficulty here is that a computer program designed to solve every problem of a certain type is not likely to solve any one of them very well. A program designed to analyze routes, flows, and so forth, on an arbitrary graph of n nodes is not very efficient for handling special types of graphs arising from geometric configurations with certain special (perhaps qualitative) features.

Perhaps a more serious problem is that the computer is not a very powerful tool for selecting logical proposals for alternative networks. It can only compare a finite number of externally proposed alternatives (the number depends on how long it takes to examine any one), whereas one would like to scan a multiparameter family of alternatives. A computer may have difficulty finding the minimum of a function of only four or five variables (to scan ten values per parameter requires 10^4 or 10^5 evaluations), particularly if the function is difficult to compute. Transportation systems involve hundreds or thousands of possible parameters. If a computer is to assist in sorting alternatives, the evaluation of any particular proposal must be done in minutes or seconds, not days.

Every computer calculation involves a sharing of labor between humans and computers. In transportation planning, the zones and the possible future networks are selected by human judgment; the computer handles calculations of the flows on a given network. Ideally computers should do those things it can do fast and humans should do those things they can do fast. The problem comes not so much in deciding who or what can do which job best but in constructing a communication link between humans and computers. Computers are very fast in performing mathematical calculations and following simple internally programmed instructions. Humans are very quick to recognize geometrical patterns and have a great capacity for switching to a new line of thought or strategy if some calculations are not progressing in the manner expected or are not giving any useful results. Human thought processes typically involve a much more complicated set of procedures than one would care to insert in a computer program, and their memories contain much more information than one would care to insert in a computer memory.

The partitioning of the study region into zones is done by hand because it would take longer to program the computer to take geometric relations into account than to do it by inspection of a map. The choice of alternative proposals is done by hand because the logic of selection involves many political and economic objectives that are difficult to quantify. A person can sift through and reject many proposals for obvious reasons, so obvious that he would not care to explain why to the computer.

The collection and analysis of data (such as O-D surveys) are shared between humans and computers because any data that are collected by hand can be punched onto cards as easily as they can be converted to any other format. Once it has the data in this form, the computer can sort, summarize, and otherwise operate on the data following simple instructions and print the results in a form convenient for interpretation. For most tasks, however, the interaction between computer and human is very awkward. It takes a long time for a person to convey information to the computer through punched cards; it also takes a long time (compared to computation times) for the computer to print out results. Consequently, in the overall division of labor most tasks are done either entirely by hand or entirely by computer.

Much of what follows is motivated by the belief that in the future more of the network analysis will be done by hand to exploit the human skills of geometric interpretation and abstract mathematical analysis, and more of the selection of alternative networks will be done by computers to exploit their capacity for numerical optimization. For the moment we are concerned mostly with the former.

There are many geometric features of real transportation systems that computer programs should be able to exploit better than at present. Any transportation network can be partitioned into parts in many different ways so that one can analyze separately the trips within parts and between parts. Whether it is advantageous to break a network into parts depends on the complexity of the trips between parts. For an "arbitrary" graph this may have no advantage because it may be just as difficult to analyze the trips between its parts separately as to analyze the original graph as a whole.

Some of the obvious ways to partition the problem are to divide the geographical region into zones (not necessarily small); to classify links according to their physical properties (freeways, main arterials, access roads, rail lines, etc.); and to classify trips according to their length, purpose, etc. Such partitions are, in fact, done in present schemes.

63   Approximations and Idealizations

The motivation for dividing a region into zones is to exploit the notion of two-dimensional distance, which the computer has difficulty recognizing. At present this notion is exploited only through the implication that all points in a zone are "close together" and approximately equivalent; thus the zone is represented by a point. Link classifications involve recognition that different links have different velocities, some roads are more important than others (some are disregarded), and some links involve particular modes, such as bus or train routes. At present trips are classified according to length, purpose, and so forth mainly as a means of constructing distribution models rather than for assignment. There are, however, other obvious qualitative features associated with these partitions that one should be able to use.

Short trips are either within a zone or between adjacent zones, and trips involve mainly minor roads. The part that is on major roads or that influences major road traffic can be analyzed locally. One need not look at the whole study region to describe the routes of short trips within a subregion.

Traffic on minor roads is not of much direct interest in transportation planning. These roads are (by definition) uncongested and are likely to remain so (or lose their designation as minor). The short trips are important mainly insofar as they contribute to the congestion on major roads. To determine the flow on major roads arising from short trips, one does not need detailed O-D data. It usually suffices to know the density of households and trips per household and to postulate that these are distributed more or less uniformly over certain regions. Perhaps one should know the location of a few centers of congestion, such as schools and shopping centers.

Because there are so many minor roads, one can probably go to the extreme of idealizing the minor road network as an arbitrarily fine grid (infinitely many roads). The zone as a whole could be idealized by a set of major roads embedded in a continuum of minor roads. It makes no difference how one idealizes the network so long as a reasonable estimate of the contribution to the flow on major roads is determined.

Anything one can do that makes sense is likely to be an improvement over what is presently done: The local street network is replaced by dummy links from a centroid to major roads plus implied links from the actual origin to the centroid. A trip that goes one block across a zone boundary is, in effect, assigned a route going to the centroid, then to a main road, another centroid, and so forth. This procedure can grossly overestimate the travel on main roads.

One might next consider intermediate-length trips, although it is not exactly clear what should be called "short" and what should be called "intermediate." Roughly speaking, intermediate-length trips are ones that cross from one zone to a neighboring zone and whose lengths are comparable with the dimension of the zone. They are likely to follow major highways for much of the route. The trip length should also be comparable with but larger than the typical spacing between major routes. The choice of major routes also depends on detailed locations of origins and destinations within the zone. It is not obvious how one might idealize this. The only feature one should obviously try to exploit is that the analysis of a given zone should involve, at most, its neighboring zones. If the study area contains a large number of zones, one should be able to analyze a few zones at a time by such decomposition. Moreover, in the analysis of these trips one can perhaps justify and even estimate errors associated with the use of a skeletal network that includes only the major roads.

Long trips are interzonal trips whose lengths are comparable with the dimensions of the study region. The simplifying feature of a long trip is that much of it is likely to be on arterial highways, freeways, or transit lines. The number of such highways or transit lines is not likely to be very large, so that they form a graph of quite manageable size. A long trip can be decomposed into the trip from origin to freeway or transit lines; the trip on the freeway or transit line; and the trip from the freeway or transit line to the destination. If a trip is on a route of minimum "distance" (travel time, cost, etc.), then the routes between intermediate points must also be optimal. This implies that the route from the origin to the freeway or transit line does not depend on the path from the freeway or transit line to the destination. To examine the flows on routes within the origin (destination) zone, it suffices to know the directions of trips (what freeways or transit lines will be used); one need not know the exact destinations (origins).

**4.2
Dense Networks**

The argument of section 4.1 suggests consideration of two extreme approximations, one in which a network with an excessively large number of links is approximated by one with a relatively small number of links, and the other in which the network is approximated by one with an infinite number of links (a continuum). The former type of approximation has been applied extensively (although its errors are seldom estimated). The latter has not been used explicitly, perhaps because the techniques of analysis are not sufficiently developed.

Square Grid

It was shown in section 3.2 that for a very fine square or rectangular grid of roads, the travel distance along the grid directions from an origin to a point with Cartesian coordinates $[x_1, x_2]$ can be represented by

$$d([0, 0], [x_1, x_2]) = |x_1| + |x_2|. \tag{4.1}$$

The *equidistance contours*, the locus of all points a distance d from the origin,

$$|x_1| + |x_2| = d, \tag{4.2}$$

are squares oriented at 45° to the grid directions, as shown in figure 4.1. This family of squares is consistent with a Huygens construction in which the contour at distance d_2 can be obtained from the contour at distance d_1 by drawing the envelope of all squares having centers along the contour d_1 and diagonal length $2(d_2 - d_1)$. This follows the procedure of section 3.3, except that the small circles are replaced by small squares.

The distance d can also be expressed in terms of polar coordinates:

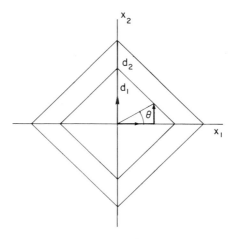

Figure 4.1
Equidistance contours for a dense
rectangular grid.

$$x_1 = R \cos \theta, \qquad x_2 = R \sin \theta,$$
$$d([0, 0], [x_1, x_2]) = R|\cos \theta| + R|\sin \theta|.$$

If one travels along any of the coordinate directions $\theta = j\pi/2$
($j = 0, 1, 2, 3$), the distance d is the same as the Euclidean
distance R, but if one travels to destinations making an angle
of 45° to the grid, $\theta = j\pi/4$ ($j = 1, 3, 5, 7$), the distance d is
$\sqrt{2} R$ or 1.41 times the Euclidean distance. If trips from the
origin $[0, 0]$ were uniformly distributed in all directions
($0 < \theta < 2\pi$) and all had Euclidean length R (destinations
uniformly distributed on a circle of radius R), then the
average d for such trips would be the same as the average
over just the positive quadrant:

$$\bar{d} = \frac{2}{\pi}R \int_0^{\pi/2} (\cos \theta + \sin \theta)d\theta = \frac{4}{\pi}R \approx 1.27\,R. \qquad (4.3)$$

Most cities have a rather complex geometry of roads, but
from most origins one can travel locally in only four
directions. In this sense, the network has properties similar
to a rectangular grid. As a general rule, the typical distance

between randomly chosen points in a city is about 1.2 times the Euclidean distance measured on a map (it is somewhat better than a perfect rectangular grid because of various extra roads making odd angles with the grid directions). One can usually obtain a sufficiently accurate estimate of the total travel distance of trips in a city simply by adding about 20 percent to the Euclidean distances.

Triangular Grid   The more directions in which one can travel from any point, the shorter will be the average travel distance between points. In the triangular grid of figure 4.2a, one can travel in six directions from any intersection. If the trip length is large compared with the grid spacing, however, the details of the local geometry of intersections is irrelevant; the distances between pairs of points in figures 4.2a and 4.2b are all nearly equal.

The local geometry of figure 4.2b avoids the objectionable six-way road intersections. With such a road system, one can devise schemes of traffic-signal coordination, with either two-way streets or alternating parallel one-way streets. Because the shortest routes between any two points (except for trips shorter than the grid spacing) would never have a 120° turn, only 60° turns, one could also prohibit such turns at all intersections. The main objection to the geometry of figure 4.2b as compared with a square grid is the peculiar shape of the land parcels, in which one would have to build triangular and hexagonal buildings.

If we shrink the mesh size to zero, then the local metric permits travel from any point in each of six directions. By symmetry, it suffices to analyze trips from the origin to points with polar coordinates R and θ $(0 < \theta < \pi/3)$, as shown in figure 4.3a.

As with the square mesh, the shortest route between two points in a triangular grid is not unique. All routes that reach the point R, θ by route segments at angles of 0 or $\pi/3$ will have the same length. If we let u and v denote the distances traveled at angles 0 and $\pi/3$, then, by the law of sines applied to the triangle of figure 4.3a,

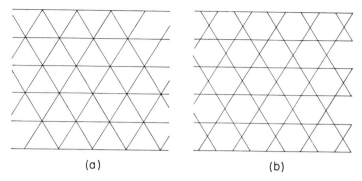

Figure 4.2
Triangular grids of roads.

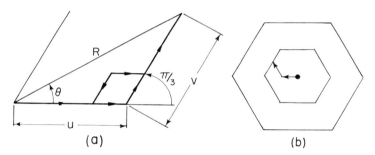

Figure 4.3
Shortest paths (a) and equidistance
contours (b) for a three-directional
grid network.

$$\frac{R}{\sin (2\pi/3)} = \frac{u}{\sin (\pi/3 - \theta)} = \frac{v}{\sin \theta}.$$

The travel distance to R, θ is, therefore,

$$d = u + v = \frac{R[\sin (\pi/3 - \theta) + \sin \theta]}{\sin (2\pi/3)}.$$

If one travels along any of the six grid directions, $\theta = j\,\pi/3$, the travel distance is the Euclidean distance R. The worst direction of travel is to a destination with $\theta = \pi/6$, for which the distance is $2R/\sqrt{3} \approx 1.15\,R$. The average travel distance to points uniformly distributed on a circle of radius R is

$$\bar{d} = \frac{2R}{\sqrt{3}} \left(\frac{3}{\pi}\right) \int_0^{\pi/3} [\sin{(\pi/3 - \theta)} + \sin{\theta}]d\theta$$

$$= \frac{2\sqrt{3}}{\pi}R \approx 1.10\ R. \tag{4.4}$$

The equidistance contours are hexagons, as in figure 4.3b. Again, one can use a Huygens construction with hexagons instead of circles. Obviously, as the number of travel directions increases, the equidistance contours approach circles.

In comparing the square and triangular nets, one should notice that the distance between points is nearly independent of the mesh size: it is unnecessary to compare the nets under the hypothesis that they have equal spacings. For example, the two could be chosen so as to have the same total length of highway per unit area, and thus nearly equal construction costs. The triangular net saves about 15 percent in travel distance, mainly at the expense of odd-shaped land parcels.

A mesh of four, five, or more two-way road directions will have still smaller average distances between points. For infinitely many directions, the metric becomes the Euclidean metric with $\bar{d} = R$; the gain from increasing two-way road directions is rather small. The worst completely connected regular road mesh, however, is one with only three travel directions along one-way roads at angles of 0, $2\pi/3$, and $4\pi/3$, for which the $\bar{d}$ over a circle of radius R is approximately $1.65\ R$.

Radial-Ring Networks In addition to the geometries described above—square grids and other multidirectional grids—analysis has been done of various networks having rotational symmetry around some point, instead of (local) translational symmetry. Considerable work has been done by various people, particularly at the Road Research Laboratory in England (Smeed, Tanner, Holroyd), on the routing of traffic over networks in a circular region. This region is intended to represent either an entire city or just the central business district of a city.

Rotationally symmetric networks all have radial roads and

circular (ring) roads. The radials are usually considered to be sufficiently numerous that they can be idealized as infinitely many radials; thus the network has complete circular symmetry. If there are only a few radials, they are distributed so as to preserve a rotational symmetry for rotations through certain angles ($2\pi/n$ if there are n radials). The number of ring roads may vary from 0 to ∞ (dense rings).

If there are no ring roads, every trip must go to the center and out again, unless the center is the destination or origin, or both origin and destination are on the same radial. If all trips have the center as origin or destination, then this is the ideal geometry. If most trips are not to the center, however, the center becomes a point of extreme (and unnecessary) congestion, and the travel distance between two points is typically rather large. If, for example, origins and destinations are uniformly distributed within a circle of radius R, then the mean travel distance is $4R/3$, about 1.47 times as large as the mean Euclidean distance, which is only about $0.90\ R$.

Even one ring road with suitable radius will relieve some congestion at the center and reduce the average travel distance considerably. If, at the other extreme, we have ring roads at every radius, the shortest routes from a point with polar coordinates $r_1, 0$ to r_2, θ are as shown in figure 4.4. If $\theta < 2$ radians and the destination has a radial distance r_2, with $r_2 < r_1$, the shortest route is along the radial road of the origin from r_1 to r_2, then along the ring road at radius r_2. If $\theta < 2$ and $r_2 > r_1$, the shortest route is along the ring road at radius r_1, then outward along the radial road at angle θ. In either case, one uses the ring road of smaller radius, r_1 or r_2. If $\theta = 2$, however, the travel distance along the ring road, $r_1\theta$ or $r_2\theta$, is equal to the distance to the center and out again along the radial at angle θ. If $\theta > 2$, the shortest route is along the radial to the center and out along the radial at θ.

Except at the origin $r = 0$, there are (two-way) roads through any point only in two perpendicular directions. The network of roads in any region having linear dimensions small compared to the distance from the center is nearly a rectangular grid oriented in the direction of the local radial

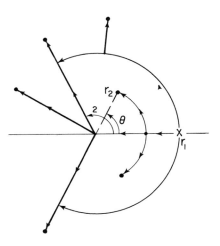

Figure 4.4
Routes on a radial-ring network.

and ring roads. For trips whose lengths are small compared to the distance to the center, the ratio of the average travel distance to the Euclidean distance is, therefore, nearly the same as that for the square grid. For trips going near or through the center, however, the travel distance is significantly less than for a square grid.

One can show that if origins and destinations are uniformly distributed within a circle defined by one of the ring roads, the average travel distance between points is very nearly the same as for a triangular grid. This efficiency in travel distance is achieved at the expense of drawing trips close to the center or through it. A certain fraction of all trips go through the center.

Almost any distortion (bending) of a square grid will result in a decrease in the average travel distance. The travel distance between opposite corners of a rectangle is independent of the path along two sides. If we bend the rectangle, however, one path is likely to become longer and the other shorter, and a trip will always choose the shorter path. Any regular grid is, in some sense, inefficient because it provides many routes of equal length between points even though a trip needs only one.

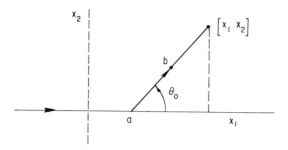

Figure 4.5
Route from a road into a
Euclidean space.

4.3
A Fast Road in a
Euclidean Space

One of the characteristic features of most transportation
networks is that they consist of a superposition of fast routes
on a much finer mesh of slower routes. This is true, for
example, of minor roads versus major roads, and of bus
routes versus transit lines. A local idealization of this might
be a single fast route embedded in a continuum of minor
routes such as those described in the last section.

Consider a single straight road superimposed on a two-
dimensional continuum of routes along which one can travel
in any direction. Suppose the velocity of travel is V on the
single fast route and v on the continuum, $V > v$. This is not
a very good representation of any road network because the
continuum can usually not be approximated with a Euclidean metric. Perhaps we should imagine a trail cut through
a jungle: one can walk slowly through the jungle in any
direction but quickly only along the trail.

Because there is no preferred direction in a continuum, no
special line from which to measure angles, we will choose a
Cartesian coordinate system $[x_1, x_2]$ with one axis parallel
with the road. Because there is no special point in the space
from which to measure the distances x_1 and x_2, it is natural
to use the road itself as the origin of the coordinate x_2, as in
figure 4.5.

Suppose, first, that the origin of a trip is on the road at
$[-\infty, 0]$, and one wishes to reach a point $[x_1, x_2]$ in the
shortest time. The optimal route must follow the road until
it reaches some point $[a, 0]$. If it leaves the road at one point,

it will not return at another point because the fastest route between two points on the road is along the road. Having left the road at $[a, 0]$, the shortest path to $[x_1, x_2]$ is a straight line making some angle θ_0 with the road. To determine the route, we need only determine θ_0 (as function of $x_1, x_2, v,$ and V).

One can immediately show that θ_0 must be independent of x_1 and x_2 for $x_2 > 0$ and a function of v/V only. It must be independent of x_1 because there is no preferred point from which to measure x_1. (The value of x_1 depends on the arbitrary choice of origin for the x_1 coordinate.) It must be independent of x_2 also because θ_0 is dimensionless, whereas x_2 has dimensions of length. The velocities v and V have dimensions of length/time; thus the value of θ_0 must depend only on the dimensionless ratio v/V.

One can also argue that the optimal route to any point b on the line from $[a, 0]$ to $[x_1, x_2]$ in figure 4.5 must follow the optimal route to $[x_1, x_2]$ as far as point b; for if there were a faster way to reach b, one would have used it to reach $[x_1, x_2]$ by way of b. Thus θ_0 is independent of the location of the destination along the line, that is, x_2.

The actual evaluation of θ_0 is a simple exercise, but even simple exercises have either "brute force" or elegant methods of solution. We consider three methods.

Direct Approach   The most direct method for evaluating θ_0 would be to evaluate the travel time as a function of θ, but with the origin at $[-\infty, 0]$ the total travel time is infinite by any route. Instead, we compare the travel time for an arbitrary θ with that for some reference value, such as $\pi/2$.

To travel from the point $[a, 0]$ to $[x_1, x_2]$ by way of a path $[a, 0]$ to $[x_1, 0]$ to $[x_1, x_2]$ requires a time

$$\frac{x_2}{v} + \frac{x_2}{V} \operatorname{ctn} \theta = \frac{x_2}{v}\left(1 + \frac{v}{V} \operatorname{ctn} \theta\right).$$

To travel along a straight line requires a time $x_2/(v \sin \theta)$. The saving is

$$\frac{x_2}{v}\left(1 + \frac{v}{V} \operatorname{ctn} \theta - \frac{1}{\sin \theta}\right).$$

A graph of this expression as a function of θ shows that it has a single maximum (if $v/V < 1$). The location of the maximum does not depend on the units for measuring benefit and, therefore, is independent of x_2 and v, which appear only as scaling factors. The maximum occurs where the derivative vanishes;

$$\frac{x_2}{v}\left(-\frac{v}{V\sin^2\theta_0} + \frac{\cos\theta_0}{\sin^2\theta_0}\right) = 0 \quad \text{or} \quad \cos\theta_0 = v/V.$$

An Infinitesimal Argument

Sometimes it is easier to determine the derivative of an objective function directly than to evaluate the objective function and then differentiate it.

Suppose, as shown in figure 4.6, one compares a path leaving the road at an angle θ with another path that remains on the road for an arbitrarily small distance ε and then leaves at a slightly larger angle. The path from $[-\infty, 0]$ to $[a, 0]$ is common to both. If we draw a circle with center at $[x_1, x_2]$ through $[a + \varepsilon, 0]$, it will cut equal lengths from the two paths to $[x_1, x_2]$. For sufficiently small ε, the circular arc between the two paths is nearly a straight line perpendicular to the paths.

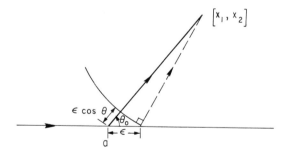

Figure 4.6
Infinitesimal displacement of a path.

From figure 4.6 one can see immediately that the path from $[a, 0]$ to $[x_1, x_2]$ travels an extra distance $\varepsilon \cos\theta$ at velocity v, whereas the path to $[a + \varepsilon, 0]$ travels an extra distance ε at velocity V. It will be advantageous for the trip to remain on the road if, for $\varepsilon > 0$, the travel time of the latter segment is smaller; that is, if

$$\frac{\varepsilon}{V} < \frac{\varepsilon \cos \theta}{v} \quad \text{or} \quad \theta < \cos^{-1}(v/V).$$

It will be preferable to leave at $[a, 0]$ if $\theta > \cos^{-1}(v/V)$. Thus if $\theta < \cos^{-1}(v/V)$, it is advantageous to increase θ, but if $\theta > \cos^{-1}(v/V)$, it is better to leave earlier ($\varepsilon < 0$) and decrease θ. The optimal angle is $\theta_0 = \cos^{-1}(v/V)$.

Huygens's Construction

In the two-dimensional continuum with a Euclidean metric, the equi–travel time contours from any origin (in this case $[-\infty, 0]$) must be everywhere perpendicular to the direction of travel. (In geometric optics, the rays are everywhere perpendicular to the wave front.) In the present situation, the direction of travel in the continuum to any point $[x_1, x_2]$ must be at angle θ_0, independent of the location of the point $[x_1, x_2]$. The equi–travel time contours must therefore make an angle of $\pi/2 - \theta_0$ with the road.

From figure 4.7 we see that the equi–travel time contour that can be reached in time t from point a can be reached either by traveling a distance vt in the continuum at angle θ_0 or by traveling a distance Vt along the road (or appropriate combinations thereof). From the geometry of figure 4.7 we see immediately that

$$\cos \theta_0 = vt/Vt = v/V. \tag{4.5}$$

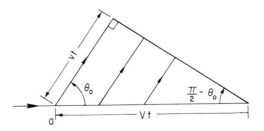

Figure 4.7
Equi–travel time contours.

In this analysis, it has been assumed that $x_2 \geq 0$. The corresponding paths for $x_2 < 0$ can be obtained by a reflection across the line $x_2 = 0$. A path leaves the road at the same angle θ_0 but is measured clockwise instead of counterclockwise from the road.

The exercise of determining an optimal angle θ_0 at which a trip should leave a road may seem a bit artificial in the context of transportation networks because, for any real network of slow-speed roads, one can travel only in certain directions. Although a traveler may not have the option of traveling in any direction he wishes, a planner may have the option of choosing any orientation between a network of slow-speed roads (velocity v) and a high-speed road (velocity V). If travel time is an important consideration, he may design the roads to meet at the angle $\theta_0 = \cos^{-1}(v/V)$.

Suppose, for example, that one were building a new suburban community on undeveloped land and wanted to connect it with a city (at $[-\infty, 0]$) by a single high-speed, high-volume route (a railway or freeway). Homes are to be built along parallel roads that can have any orientation relative to the high-speed route. There are no trips between homes within this suburb, only from home to the city or from the city to home. If travel time were the only issue, one would build the roads at an angle θ_0 to the main route, as shown in figure 4.8. Because there are no trips between homes, there is no need to build roads perpendicular to those at angle θ_0, and there is no reason why the roads above the main road should be continuations of those below.

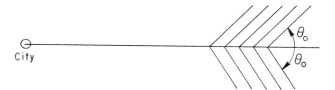

City

Figure 4.8
Road network for a suburban
community having trips only to or
from a city.

Although travel time is considered the objective in this optimization problem, there are many other problems that are mathematically equivalent to this but have somewhat different interpretations. Their main feature is one cost per unit distance on one route and another on a second route; these routes must join so as to minimize the total cost. The cost could be construction cost (per unit length) or construction plus travel cost. In all such problems, there is an optimal angle at which the routes join, and this angle depends on the ratio of the costs per unit distance of the two routes.

Returning to our problem of determining the fastest route between two points, we consider the fastest route from a point on the road at $[y_1, 0]$ (y_1 finite) to $[x_1, x_2]$. Because the properties of our metric are invariant with respect to translations of coordinates in the direction of the road, we may, without loss in generality, assume that $y_1 = 0$; that is, we choose the origin of the trip as the origin of the coordinate system. It suffices to analyze the special case $x_1 > 0$, $x_2 > 0$ because the geometry of this problem is invariant with respect to reflections across either coordinate axis. Any optimal path from $[0, 0]$ to $[x_1, x_2]$ has a reflection across the vertical axis that yields the optimal path from $[0, 0]$ to $[-x_1, x_2]$ and a reflection across the horizontal axis that yields the optimal path from $[0, 0]$ to $[x_1, -x_2]$.

If $\tan^{-1}(x_2/x_1) < \theta_0$, it is possible to reach the destination by travel in the positive direction along the road and leaving the road at an angle θ_0. It follows from analysis of figure 4.6 that one would not leave the road at any smaller angle. If $\tan^{-1}(x_2/x_1) > \theta_0$, then to leave the road at an angle θ_0 would necessitate traveling in the negative direction along the road. The travel time on the road, however, is proportional to the absolute value of the coordinate displacement. The optimal place to leave the road is not found by differentiation of some travel time function. The optimal place to leave is at the origin (where the travel time function has a discontinuous derivative, a cusp).

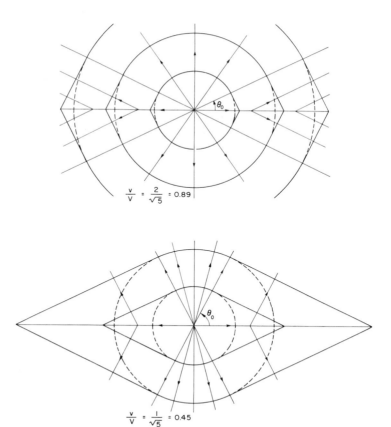

Figure 4.9
Equi–travel time contours and
routes from a point on a road to
points in a Euclidean space.

Some fastest routes and equi–travel time contours are
shown in figure 4.9 for $v/V = 0.89$ and $v/V = 0.45$. For
$\tan^{-1}(x_2/x_1) > \theta_0$ the contours are circles; for $\tan^{-1}(x_2/x_1)$
$< \theta_0$ they are straight lines; at angle θ_0 the line is tangent to
the circle. The broken lines show the continuation of the
circle, the contours that would exist in the absence of the
road.

Finally, we wish to determine the fastest route from an
arbitrary point $[\,y_1, y_2\,]$ to $[\,x_1, x_2\,]$, of which the previous
examples are all special cases. Although this is not a difficult

problem, we are approaching it as if it were in order to illustrate the sort of graded approach one should take in analyzing a complicated problem with many parameters. The idea is to postpone for as long as possible any attempt to solve the problem explicitly. One should first try to solve simple examples and eliminate as many parameters as possible.

In an abstract sense, to find the fastest route between any two points in a continuum requires the determination of a family of curves, one curve for each O-D pair; that is, we seek relations of the form

$$f(z_1, z_2; x_1, x_2, y_1, y_2) = 0.$$

For fixed x_1, x_2, y_1, y_2, this is the equation of a curve in the space with coordinates z_1, z_2. The curve is constrained to pass through the points $[x_1, x_2]$ and $[y_1, y_2]$. Although we shall not actually express the equations for the route in this form, we shall describe enough features of the relation f so that we could determine it.

The route must, first of all, be a piecewise linear curve because the shortest path between points in the continuum is a straight line. The route must, furthermore, be either a straight line from $[y_1, y_2]$ to $[x_1, x_2]$ or a route consisting of a straight-line path to the road, a segment along the road, and a straight-line path to $[x_1, x_2]$. The only place the route can change directions is where it joins the road. Again one can use various symmetries to eliminate the dependence of the route on some of the parameters y_1, y_2, x_1, x_2.

Translation Symmetry   There is no preferred position for the origin of the coordinate system. For any x_1, x_2, y_1, y_2, the fastest route from $[y_1, y_2]$ to $[x_1, x_2]$ is a translation of the route from $[0, y_2]$ to $[x_1 - y_1, x_2]$. It suffices, therefore, to consider only the case $y_1 = 0$. This eliminates one parameter from the relation f.

Reflection Symmetry   Because all routes are two-way, the geometry is invariant with respect to reflections across either the vertical or the horizontal axis. If one knows the fastest route from $[0, y_2]$ to $[x_1, x_2]$, then, by reflection, one also knows the fastest route from $[0, -y_2]$ to $[x_1, -x_2]$ and from $[0, y_2]$ to

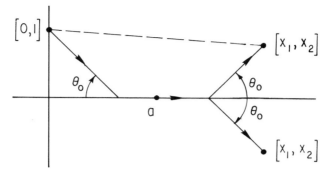

Figure 4.10
Fast routes from $[0,1]$ to $[x_1, x_2]$.

$[-x_1, x_2]$. It suffices, therefore, to consider only the fastest route from $[0, y_2]$ to $[x_1, x_2]$ with $y_2 > 0$ and $x_1 > 0$ (but all values of x_2, both positive and negative).

Scaling Symmetry Because there is no spacing between streets, distance between interchanges, and so forth, there is no natural unit of distance, time, or cost. Given the fastest route from $[0, y_2]$ to $[x_1, x_2]$, then, for every positive number b, the fastest route from $[0, by_2]$ to $[bx_1, bx_2]$ has exactly the same geometry, differing only in that the units of distance are changed or, equivalently, the scale of the graph is changed. If one can find the fastest route from $[0, 1]$ to every point $[x_1, x_2]$, then one will also know, by rescaling of coordinates, the fastest route from $[0, y_2]$ to $[x_1, x_2]$ for any $y_2 > 0$ (in effect, one uses y_2 as the unit of length). This eliminates a second parameter from f.

From the analysis of the special case $y_2 = 0$, we can easily identify the nature of the fastest route. If a fastest route from $[0, 1]$ to $[x_1, x_2]$ travels along the road anywhere and passes through some point $[a, 0]$ for some value of a, as in figure 4.10, then the path from $[0, 1]$ to $[a, 0]$ and the path from $[a, 0]$ to $[x_1, x_2]$ must be optimal. If the optimal path actually travels along the road (not just crosses it or touches it at one point), then the path must approach the road at an angle θ_0 and leave it at an angle θ_0. Because this uniquely determines the path to and from the road, it also uniquely determines the optimal path from $[0, 1]$ to $[x_1, x_2]$, provided the path follows the road.

The only other possible candidate is the optimal path that does not follow the road, that is, the straight-line path from $[0, 1]$ to $[x_1, x_2]$. Because there are only two potential candidates for the fastest route, it is not very difficult to evaluate the travel time between two points along the two routes and choose the smaller.

This last step can be done by the geometric construction shown in figure 4.11. First draw the (broken) lines from $[0, 1]$, making angles θ_0 with the road, and continue them past the road. To reach any point in the cone between these lines, one should use a direct route, even if it crosses the road. Draw equi–travel time contours (circles) in this region.

To reach a point just to the right of point 1, one would approach the road at an angle θ_0, follow the road a short distance, and then leave at an angle θ_0 again. The equi–travel time contour passing through point 1 is a straight line to the right of point 1, tangent to the circle and perpendicular to the broken line. The line extends until it reaches the road.

To reach a point just above the road, one would leave the road at an angle θ_0 above the road. The equi–travel time contour is symmetric, with the road as an exis of symmetry. At point 2, where the circle through point 1 intersects the straight-line contour, the travel time is again the same by way of the road or by a direct path. To the left of point 2, it is faster to travel by the driect path. The locus of points like 2 generated in this way form a curve (broken line) that, along with the broken line through point 1, defines the region of destinations for which the road would be used. The curve is actually a parabola with an axis along the other broken line through $[0, 1]$ at angle $+\theta_0$.

4.4
A Fast Road on a Square Grid

We considered a fast route on a Euclidean metric as a first example because the geometry depends only on one parameter, v/V. If we build a fast route over a square grid (or some other grid lacking rotational symmetry), the route pattern will also depend on the orientation of the fast route relative to the grid.

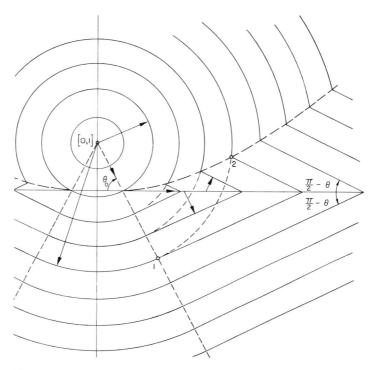

Figure 4.11
Equi–travel time contours from a
point [0,1].

Because of symmetry, one would guess that the "best" and
"worst" orientations of a fast route on a square grid are
either parallel or diagonal to the grid. The argument for this
depends, of course, on the objective, but these two orientations are special in the sense that there are only two axes
and two diagonals. If the fast route were placed at some
angle α, then the orientations α, $-\alpha$, $\pi/2 + \alpha$, and $\pi/2 - \alpha$
would produce similar patterns. If the objective has the same
values for α, $-\alpha$, $\pi/2 + \alpha$, and $\pi/2 - \alpha$, and is differentiable
with respect to α, then the derivative will vanish at $\alpha = 0$
and $\pi/4$; in any case, $\alpha = 0$ and $\pi/4$ must be at least local
minima or maxima of the objective function.

If the fast road is to have the same speed as the grid, $V = v$, then one should certainly build the road at $\alpha = \pi/4$. Because there is already a dense grid, nothing is gained by adding one more road along one of the grid directions; but if it is built at $\pi/4$, some trips will benefit from having another possible direction of travel. Something is gained even if the speed on the new road is less than on the grid but larger than $v/\sqrt{2}$, so that it is faster to travel a diagonal at speed V than two sides of a square at speed v. If $V/v > 1$, however, it is not obvious which orientation is better.

The methods for determining the fastest routes are the obvious generalization of the methods described in section 4.3 for a fast road in a Euclidean space. The translation, reflection, and scaling symmetries are retained. The only difference is that a path cannot join or leave the fast route at any angle θ; it can do so only at $\theta = \pm\pi/2$ if the road is parallel with the grid or at $\theta = \pm\pi/4$, $\pm3\pi/4$ if $\alpha = \pi/4$. Also, the shortest allowed path in the continuum is not a straight line; it consists of trip legs along the grid directions with no reversal of direction, as described in section 4.2.

Suppose we again choose a coordinate system with a fast road along the axis $x_2 = 0$ (the slow roads at $\alpha = 0$ and $\pi/2$ or at $\pi/4$ and $3\pi/4$). If a trip originates on the fast road at $[0, 0]$, the fastest routes and equi−travel time contours are as shown in figure 4.12 for $V/v = 1$, $3/2$, and ∞.

For $V = v$, and the "fast" road parallel with the grid as shown in 4.12a, the extra road has no effect; the equi−travel time contours are squares as in figure 4.1. But, for the extra road at $\alpha = \pi/4$ as in 4.12b, the new road offers another direction of travel that will be exploited if the destination is at an angle $|\theta| < \pi/4$. For comparison, the broken curve (circle) shows the contours for a travel time of 2 and a Euclidean metric, which corresponds to figure 4.9 (but with $V/v = 1$, $\cos\theta_0 = v/V$ yields $\theta_0 = 0$). In 4.12b, the contours come closer to the circles.

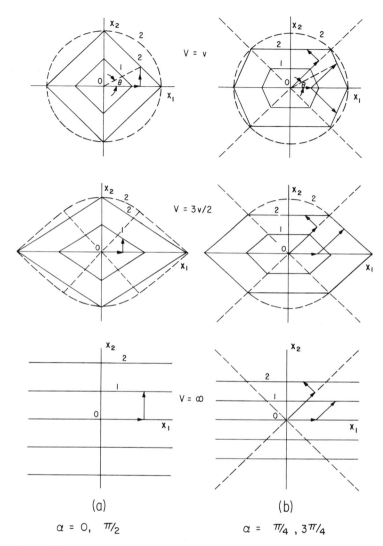

Figure 4.12
Equi–travel time contours for a
fast road on a square grid; trip
origin on the fast road.

For $V/v > 1$, as illustrated in figure 4.12 for $V/v = 3/2$, all trips in 4.12a will use the fast road for the horizontal leg of the trip instead of the horizontal grid roads; all trips (except those at $\theta = \pi/2$) will benefit from the fast road. In 4.12b only trips with $\theta < \pi/4$ will benefit from the fast road because they must leave the road at an angle of $\pi/4$.

The broken-line curves again show the corresponding contours for a Euclidean metric. The angle $\cos \theta_0 = v/V = 2/3$, at which the circular contours meet the linear contours as in figure 4.9, is sufficiently close to $\pi/4$ that in 4.12b one cannot distinguish between the solid-line contour for $\theta < \pi/4$ and the corresponding broken-line contour for the Euclidean metric. This illustrates the fact that in figure 4.9 it is not important to leave the fast road at exactly the optimal angle θ_0 because, at that angle the travel time is insensitive to small errors in route direction.

In the extreme case of $V/v = \infty$, both 4.12a and 4.12b give horizontal equi–travel time contours. Case 4.12a is preferred, however, because the travel time to any point $[x_1, x_2]$ is $|x_2|/v$, whereas in 4.12b it is $\sqrt{2}|x_2|/v$ (the vertical leg must be traversed at an angle of $\pi/4$). For a Euclidean metric, $\cos \theta_0 = v/V = 0$ or $\theta_0 = \pi/2$; the contours are as in 4.12a. For $V/v = \infty$ the total trip time is access time from the fast road to the destination. To minimize trip time, one must minimize the distance from a line to a point; one should leave the line at right angles.

For any finite value of $V/v > 1$, a fast road oriented parallel with the grid has the advantage (as compared with an orientation of $\alpha = \pi/4$) of better access from road to destination and a benefit to users leaving in any direction θ. The road with orientation at $\alpha = \pi/4$ has the advantage of providing an extra direction of travel and fast travel in a direction that, in the absence of the new road, was relatively slow. Which orientation yields the larger total benefit depends on the value of V/v and the angular distribution of the destinations (but not on the trip length because there is no natural unit of length).

To illustrate how the benefit depends on V/v and the angular distribution of trips, we will compare the travel time saving for the two orientations for a uniform distribution of destinations over either a circle or a square. The saving in travel time of a trip from $[0, 0]$ to $[x_1, x_2]$, $x_1, x_2 > 0$ resulting from use of the fast road is

$$x_1(v^{-1} - V^{-1})$$

in 4.12a and

$$\begin{cases} (x_1 - x_2)(\sqrt{2}v^{-1} - V^{-1}) \text{ for } x_1 > x_2, \\ 0 \text{ for } x_2 > x_1 \end{cases}$$

in 4.12b. If trips are uniformly distributed on a circle of radius R, the average benefit per trip is

$$R\left(\frac{1}{v} - \frac{1}{V}\right)\frac{2}{\pi} \int_0^{\pi/2} \cos \theta \, d\theta = \frac{2R}{\pi}\left(\frac{1}{v} - \frac{1}{V}\right)$$

in 4.12a and

$$R\left(\frac{\sqrt{2}}{v} - \frac{1}{V}\right)\frac{2}{\pi} \int_0^{\pi/4} (\cos \theta - \sin \theta) d\theta$$

$$= \frac{2R}{\pi}\left(\frac{\sqrt{2}}{v} - \frac{1}{V}\right)(\sqrt{2} - 1)$$

in 4.12b. (If there is a distribution of trip lengths R, we need only replace R by its average value.) The benefit from 4.12a exceeds 4.12b if

$$\frac{1}{v} - \frac{1}{V} > \left(\frac{\sqrt{2}}{v} - \frac{1}{V}\right)(\sqrt{2} - 1)$$

or, equivalently, if

$$V/v > \sqrt{2} \approx 1.41. \tag{4.6}$$

If, before the fast road was built, trips were uniformly distributed over the equi–travel time contour (a square) at travel distance R, the average benefit per trip would be

$$\left(\frac{1}{v} - \frac{1}{V}\right)\frac{1}{R} \int_0^R x_1 dx_1 = \left(\frac{1}{v} - \frac{1}{V}\right)\frac{R}{2}$$

in 4.12a and

$$\left(\frac{\sqrt{2}}{v} - \frac{1}{V}\right)\frac{1}{\sqrt{2}R}\int_0^{R/\sqrt{2}}\left(\frac{R}{\sqrt{2}} - x_2\right)dx_2 = \left(\frac{\sqrt{2}}{v} - \frac{1}{V}\right)\frac{R}{4\sqrt{2}}$$

in 4.12b. For this distribution, 4.12a is better than 4.12b if

$$V/v > 2 - 1/\sqrt{2} \approx 1.29. \tag{4.7}$$

The uniform distribution over a square gives less weight to the trips at $\theta = \pi/4$ than the uniform distribution over a circle, consequently decreasing the weight attached to the direction of maximum benefit for 4.12b. This is the reason why 4.12a is preferred for a larger range of velocities for the square than for the circle. In either case, however, a typical fast road is likely to have a value of V/v exceeding 1.3 or 1.4 (it is likely to be 2 or more), which means that the fast road oriented parallel to the grid is preferred, at least for trips originating on the fast road. This is also typically the direction with the lowest construction cost.

To describe the fastest route between any two points not on the fast road, it suffices to consider the fastest route from $[0, 1]$ to $[x_1, x_2]$. The routes from any other origin $[y_1, y_2]$ can be inferred from the translation, reflection, or scaling symmetries described for the Euclidean metric in section 4.3. Figure 4.13a shows the equi–travel time contours from any origin $[0, 1]$ for the fast road parallel to the grid and $V/v = 2$; figure 4.13b shows them for the fast road at $\alpha = \pi/4$. These are the analogs of figure 4.11 for the Euclidean metric. The broken-line contours are those that would replace the contours labeled 2 and 5 if the continuum had a Euclidean metric.

These contours can be constructed in a manner similar to the ones in figure 4.11. In figure 4.13a, for short trips from the point $[0, 1]$, the equi–travel time contours are squares (instead of circles), at least until the trip can reach the fast road. If the destination is below the fast road, the fastest route is a vertical leg to the fast road from $[0, 1]$ to $[0, 0]$, a horizontal leg along the fast road from $[0, 0]$ to $[x_1, 0]$, and then a vertical leg to $[x_1, x_2]$. The contours below the fast road are similar to those in figure 4.12, except for the labels

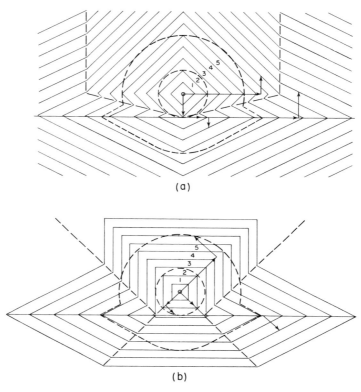

(a)

(b)

Figure 4.13
Equi–travel time contours for a
fast road superimposed on a
square grid. The fast road is
parallel with the grid in (a), 45° to
the grid in (b).

on the contours, because all such trips pass through [0, 0] and all points below the road that are equidistant from [0, 0] are also equidistant from [0, 1].

One can extend these contours above the fast road by symmetry until they meet the corresponding square contours for the direct route from [0, 1]. The shed boundary between the two routes, shown by the broken lines, is the analog of the parabolic curve of figure 4.11.

The equations for the shed boundary are quite simple. If the height of the destination $[x_1, x_2]$ is above the road but below the origin—that is, $0 \leq x_2 \leq y_2$ as in figure 4.14a—

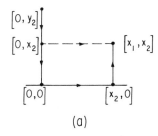 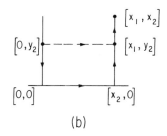

(a)   (b)

Figure 4.14
Optimal routes from $[0, y_2]$ to
$[x_1, x_2]$; fast road parallel with
the grid.

then both of the two possible optimal route types could be chosen to go first to the point $[0, x_2]$, then either by a direct horizontal leg to $[x_1, x_2]$ or by a route from $[0, x_2]$ to the fast road and then up. The latter route is preferred if the saving of time on the fast route is sufficient to compensate for the extra trip length, $2x_2$, required to reach the fast road and return. Note that this choice is independent of y_2 as long as $0 \leq x_2 \leq y_2$ because it need not be made until the routes separate at $[0, x_2]$. The fast road is preferred if

$$\frac{2x_2}{v} + \frac{|x_1|}{V} < \frac{|x_1|}{v},$$

so that

$$x_2 < \frac{1}{2}\left(1 - \frac{v}{V}\right)|x_1|.$$

Thus there is a boundary line at slope $\pm(1 - v/V)/2$ between points to which the fastest route goes via the grid and those to which the fastest route goes via the fast road.

If the destination is farther away from the fast road than the origin, however—$0 \leq y_2 \leq x_2$ as in figure 4.14b—then, in effect, the role of origin and destination are reversed in the last argument. A trip from $[0, y_2]$ to $[x_1, x_2]$ will go by the fastest route to $[x_1, y_2]$, independent of x_2 (for $y_2 \leq x_2$). The fast road is used if

$$y_2 < \frac{1}{2}\left(1 - \frac{v}{V}\right)|x_1|.$$

For a given origin $[0, y_2]$, the shed boundary between the routes is a vertical line $|x_1| = $ constant.

In figure 4.13a we have chosen an arbitrary value of y_2. (If the contour labels are interpreted as the travel times with $v = 1$, then the value of y_2 is 2.) For any other value of y_2, the whole figure is scaled proportional to y_2. For any given y_2, the fast road is used by all trips except for a vertical strip enclosed by the broken lines. The larger y_2, the wider is the strip.

In figure 4.13b, for the fast road at $\alpha = \pi/4$, the contours for short trips are again squares but with axes parallel to the fast road. The contours of figure 4.13b with the label $2\sqrt{2}$ would coincide with the fast road. The only trips with $x_1 > 0$ that would use the fast road are those that could reach it at $[y_2, 0]$, still travel along it in the positive direction, and still leave it at an angle of $\pm\pi/4$. The shed boundaries between trips traveling via the fast road or directly are 45° lines from the points $[\pm y_2, 0]$.

To compare further the benefits of a fast road parallel to or at an angle $\alpha = \pi/4$ to the grid, we might imagine that trips can originate anywhere in the plane with a uniform density ρ and that, from any origin $[y_1, y_2]$, the destinations are uniformly distributed over a square at travel distance R from the origin. Because the benefit for an individual trip is independent of y_1, there exists a benefit per unit length of road that is equal to ρ times the benefit to trips originating at $y_1 = 0$ with a density of one trip per unit length of y_2.

To evaluate the benefit of the fast road parallel to the grid, one must evaluate the average benefit per trip from $[0, y_2]$ to points uniformly distributed on a square as in figure 4.15a and 4.15b, multiply by ρ, and integrate with respect to y_2. One must consider separately the cases $|y_2| < (1 - v/V)R/2$, as in figure 4.15a, for which there are three types of trips, illustrated as 1, 2, and 3, and $(1 - v/V)R/2 < |y_2| < R$, for which there are only two types of trips, shown as 1 and 2. The result of the integration to obtain the total benefit is:

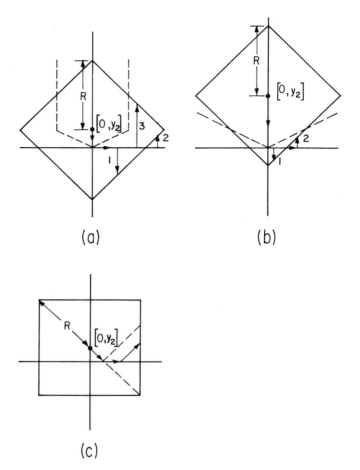

Figure 4.15
Benefits for a fast road on a
square grid.

Benefit per unit length $= \dfrac{\rho R^2}{v6}\left(1 - \dfrac{v}{V}\right)\left(2 - \dfrac{v}{V}\right).$ (4.8)

If the fast road is at $\alpha = \pi/4$, the integration is over the square of figure 4.15c. This integration yields:

Benefit per unit length $= \dfrac{\rho R^2}{v6\sqrt{2}}\left(1 - \dfrac{v}{V\sqrt{2}}\right).$ (4.9)

If there were a distribution of trip length R, the benefit per unit length would be the same as in (4.8) and (4.9), except that R^2 would be interpreted as the average square trip length. The simple dependence of (4.8) and (4.9) on R results from the fact that there is no natural unit of length. For dimensional reasons, the benefit must be proportional to $\rho R^2/v$. The benefit per unit length does, however, depend on the angular distribution of the trips in a complicated way.

Our main interest in (4.8) and (4.9) is to assess their dependence on v/V. As expected, (4.8) vanishes for $v/V = 1$; a road at $\alpha = \pi/4$ is preferred if V is equal to or slightly larger than v. If, however, V is sufficiently large, the fast road should be built parallel with the grid. Specifically, (4.8) exceeds (4.9) for

$$V/v \geq 1.37, \tag{4.10}$$

a result quite similar to (4.6) and (4.7).

**4.5
Limited-Access
Fast Road**

In many transportation systems having a fast route super-imposed on a grid, one can enter or leave the fast route only at certain points (freeway interchanges, train stations). The interchanges might, for example, be equally spaced a distance D apart, with D large compared with the spacings on the slow-speed grid. One can still idealize this by superimposing the fast road on a continuum.

The analysis of the fastest routes between points $[y_1, y_2]$ and $[x_1, x_2]$ is more complex than in section 4.4 because the discreteness of the access points destroys most of the symmetry schemes that were exploited in the case of unlimited access. One must now consider the position of the origin $[y_1, y_2]$ relative to the nearest access points. Because the distance to the nearest access point to the east may be different from that to the west, the east-west (E-W) reflection symmetry is lost. The E-W translational symmetry is lost except for translations by multiples of D, and the scaling symmetry is lost because origin and destination coordinates must be considered relative to the length D. The fastest route from $[y_1, y_2]$ to $[x_1, x_2]$ is now really a function of all four variables y_1, y_2, x_1, and x_2.

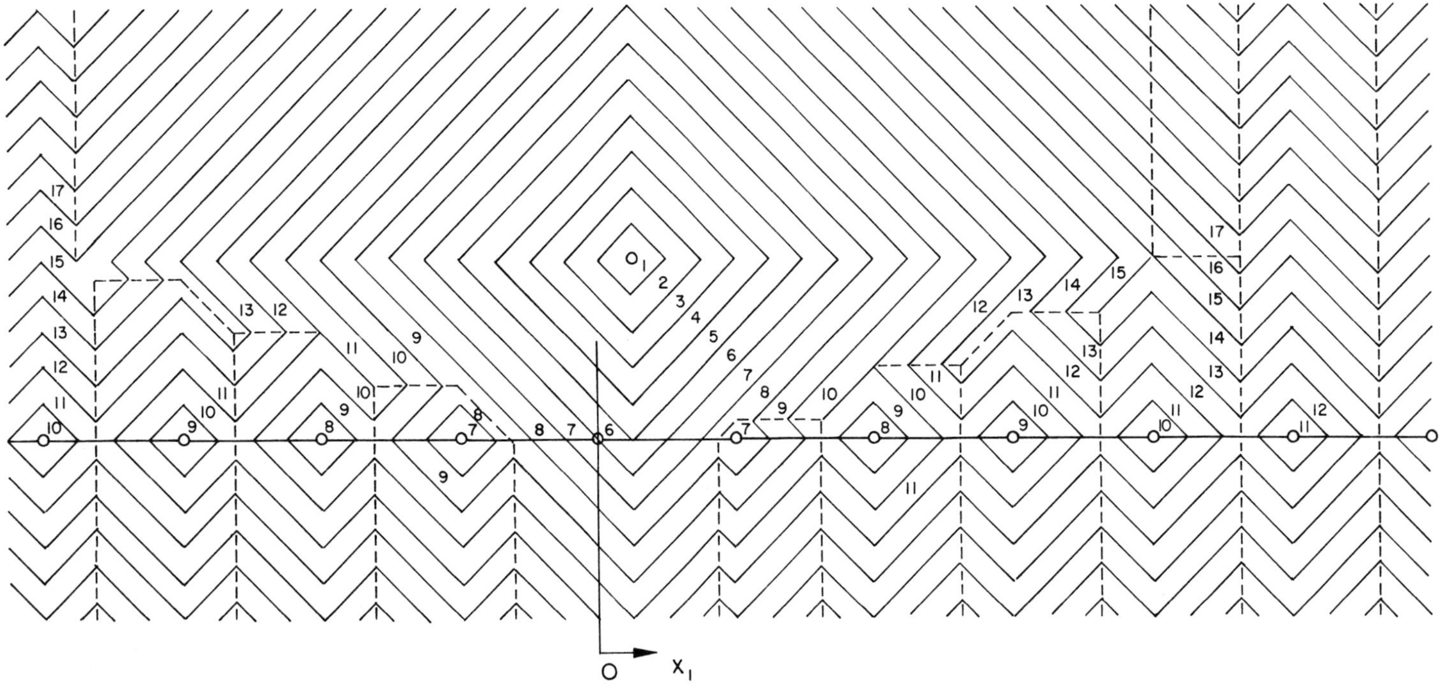

Figure 4.16
Equi–travel time contours for a limited access fast road on a square grid.

Figure 4.16 illustrates a possible family of equi–travel time contours for a limited-access fast road superimposed parallel to one of the travel directions of a fine rectangular grid (a limited-access analog to figure 4.13a). In this example, $V = 4v$, the distance of the trip origin from the fast road is comparable with D (actually $y_2 = 5D/4$), and the distance y_1 from the nearest westward access point at $[0, 0]$ is less than the distance to the nearest eastward access point at $[0, D]$.

The contours near the trip origin are squares; they remain so at least until the trip can reach the nearest access point at $[0, 0]$ (in six time units in figure 4.16). From this point a trip can reach the next interchange to the west or east, $[-D, 0]$ or $[+D, 0]$, in seven time units by traveling south to the fast road (five time units), backtracking to $[0, 0]$, and then proceeding east to $[D, 0]$ by way of the fast road. In this example, all trips that use the fast road will enter at $[0, 0]$ regardless of travel direction. If the origin were somewhat nearer the point $[D, 0]$, the eastbound trips would enter at $[D, 0]$, the westbound trips at $[0, 0]$. Specifically, an eastbound trip will enter the fast road at $[D, 0]$ if

$$\frac{D - y_1}{v} < \frac{y_1}{v} + \frac{D}{V},$$

so that

$$y_1 > \frac{D}{2}\left(1 - \frac{v}{V}\right). \tag{4.11}$$

For $v = V$, the eastbound trip would always enter at $[D, 0]$ for any $y_1 \geq 0$. For $V = \infty$, $v/V = 0$, it goes to the nearest access point; that is, it goes to $[D, 0]$ if $D/2 < y_1 < D$.

A trip to $[x_1, x_2]$ that uses the fast road will leave the fast road at the nearest exit either to the east or west of the point $[x_1, 0]$. If we write

$$x_1 = jD + x, \qquad j = \text{integer}, \quad 0 \leq x < D,$$

the trip leaves at $[jD, 0]$ or $[(j + 1)D, 0]$. It may next proceed to the point $[x_1, 0]$ before traveling north or south to $[x_1, x_2]$. The shed boundary between the exit points at jD or $(j+1)D$ satisfies a relation similar to (4.11). A trip exits at jD if

$$x < \frac{D}{2}\left(1 + \frac{v}{V}\right) \quad \text{or} \quad D - x > \frac{D}{2}\left(1 - \frac{v}{V}\right). \tag{4.12}$$

In figure 4.16, for $v/V = 1/4$, the vertical shed lines shown by the broken lines are at $x = (D/2)(1 + 1/4) = 5D/8$. The exit point depends on x_1 but not on x_2.

We still must compare the trip time by way of the fast road with the trip time by way of the grid. All trips with $x_2 < 0$ (that is, to destinations south of the fast road) will use the fast road, except possibly for some trips with an E-W trip component that is too small, $|x_1| < D$. Although the contours with $x_2 < 0$ have a very irregular shape, one can see that, on a scale of distance large compared with D, the contours wiggle around straight lines of slope V/v, similar to figure 4.13a.

Trips with $x_2 > 0$ must backtrack to reach the fast road. The shapes of the contours and shed lines are sensitive to the value of y_1/D. If D were small compared with y_1 and we drew figure 4.16 on a scale relative to y_1, it would look very similar to figure 4.13a. It would approach such a figure for $D/y_1 \to 0$, that is, $D \to 0$. At the other extreme, $y_1/D \to 0$, nearly all trips use the fast road, and the pattern of contours above the fast road looks the same as below. The region of square contours between the vertical shed boundaries, which in figure 4.16 covers most of the figure, shrinks to a narrow strip.

4.6 Comments These examples begin to show some of the types of transportation issues that can be analyzed more easily by analytic methods than by computers. The advantage of an analytic approach diminishes rapidly as the complexity of the formulas increases. If one must resort to numerical evaluation of the formulas in order to study their properties, one might as well do the analysis numerically from the start. In an analytic approach, one typically sacrifices accuracy in order to demonstrate conveniently how the solution of some (idealized) problem depends on a number of parameters. A simple formula showing the interrelation among several variables is certainly more useful that a multidimensional tabulation of numbers.

Analytic procedures should be most helpful in the final phase of transportation planning procedures, the network evaluation. Here one wishes to know how the behavior of traffic depends on the characteristics of the network. There are many parameters involved in describing a network. If any part of this evaluation can be done analytically, it would at least reduce the space of independent variables that must be investigated numerically.

Most of the issues we have discussed are ones we would not wish to analyze by the lengthy procedures outlined in chapter 1. Although one could, perhaps, more accurately compare nearly square grid networks with a transportation system having more travel directions by going through the complete network evaluation procedure, it would be extremely expensive and would not yield results much different from those described by a simple idealized geometry. The advantages of some (nearly) radial geometry are also more clearly demonstrated by an idealized model than by a "realistic" one.

Questions of how and where to build a high-speed highway or transit line involve many issues, not all of which one would care to investigate by long computations. The idealized example of section 4.4 shows clearly that access is typically a more important issue than creating a new travel direction. Several people have also investigated the effect of superimposing a high-speed ring road on an existing radial network. The example of section 4.5 only hints at how one might attack a number of problems relating to the spacing between transit stations, bus stops, or freeway interchanges. This type of analysis can be used to evaluate the increased access travel resulting from larger spacing between access points. Ultimately this could be balanced against the benefits on the fast route resulting from larger spacing, namely higher speed due to fewer stops for a transit or bus line or less congestion for a freeway. Choice of an "optimal" spacing should clearly not be very sensitive to the detailed geometry of the underlying network. One would not do a complete transportation evaluation to determine the spacing between access points.

As yet, little has been done to exploit analytical techniques in the transportation planning process. This may be partly because the analytic techniques have not been developed far enough to show the errors induced by applying them to nonidealized situations, but perhaps also because little has been done on the network evaluation phase by any method. This is where the analytic techniques could be most useful, since the numerical techniques have, for the most part, failed to provide a mechanism for exploring a wide range of possible designs.

Problems 1

A road network permits travel from any point only in directions with angles $\theta_j = 2\pi j/n$ relative to some fixed coordinates, $n \geq 3, j = 0, 1, \ldots, n - 1$. (Note that if n is odd, roads are one-way.) What is the shortest travel distance to any point on a circle of radius R? What is the average distance if destinations are uniformly distributed on this circle?

2

A freeway with velocity V is built through a region covered by a fine two-way triangular grid of roads with velocity v (travel in six directions). The freeway runs perpendicular to one of the grid directions; that is, if the grid travel directions are $\theta = \pm 60°, \pm 120°, 0°, 180°$, then the freeway is at $\theta = \pm 90°$. For an origin on the freeway, sketch the shape of the equi–travel time contours.

3

A city has ring roads and radial roads at very close spacing. Draw equidistant contours from an arbitrary point (not the center of the city).

4

On a square grid of roads with velocity v, two freeways are built with velocity $V = 2v$. The two freeways intersect at right angles, each parallel to one of the grid axes. Draw the equi–travel time contours from an origin on one of the freeways, and the paths of shortest travel time. In drawing the contours, make the origin of trips rather close but not at the intersection of the two freeways.

5

A circular city of radius R has a fine network of radial and ring roads with velocity v. A freeway ring road with velocity $3v$ built at the edge of the city, at radius R. Find the fastest routes from a point at radius $5R/6$ to any point in the city.

6

Washington, D.C., has a system of approximately radial roads superimposed on a nearly rectangular grid. Imagine a hypothetical network with dense radial roads at all angles θ from the city center, superimposed on a fine rectangular grid of two-way streets in the directions $\theta = 0$ and $\pi/2$. Starting from a point with rectangular coordinates $[1, 0]$ (polar coordinates $r = 1, \theta = 0$), find the shortest path to any point in the plane. Sketch the shape of the equidistance contours from the point $[1, 0]$.

7

Repeat problem 4 with the freeways oriented at $\alpha = \pi/4$.

8

Repeat problem 6 with a dense system of ring roads added to the network.

9

A square grid of freeways with velocity V is superimposed on a fine square grid of city streets with velocity v. The two systems are oriented in the same way. Draw the equi–travel time contours from some origin at the intersection of two freeways to points anywhere in space. Treat the city streets, but not the freeway grid, as a continuum.

References   There is a sizable literature on travel distances and flows over idealized networks of routes, particularly for hypothetical cities of circular shape. Much of this literature is summarized in the following:

1
Holroyd, E. M. *Theoretical Average Journey Lengths in Circular Towns with Various Routing System*, RRL 43. Road Research Laboratory, 1966.

2
Smeed, R. J. "The Effect of the Design of Road Network on the Intensity of Traffic Movement in Different Parts of a Town with Special Reference to the Effect of Ring Roads." Tewksbury Symposium, University of Melbourne, 1970.

One particularly interesting type of network for a circular city is a superposition of families of intersecting logarithmic spirals:

3
Miller, A. J. "On Spiral Road Networks," *Transportation Science* 1 (1967): 109–125.

For a further discussion of the effect of high-speed ring roads, see

4
Blumenfeld, D. E., and Weiss, G. H. "Circumferential-direct Routing in a Circular City," *Transportation Research* 4 (1970): 385–389; and "Routing in a Circular City with Two Ring Roads," ibid., pp. 235–242.

5
Pearce, C. E. M. "Locating Concentric Ring Roads in a City," *Transportation Science* 8 (1974): 142–168.

5 FLOWS ON NETWORKS

5.1
Introduction

The notion that a link on a network is supposed to be used to transport something is rather basic to the whole purpose of transportation planning. The term *flow* is customarily used to refer to the amount of whatever is transported. Unfortunately, the term is used in many different ways.

Books on "network flows" (not necessarily in transportation) usually fail to make a distinction between flow as a total amount of goods transported, with units of quantity (volume, number, cost), and as the rate at which goods are transported, with units of quantity/time. It is merely stated that to each directed link (i, j) there is associated a non-negative number f_{ij}, and these numbers satisfy certain relations corresponding to the postulate that goods do not disappear and are not created except through explicitly specified "sources" or "sinks." Whereas distances are basically defined on pairs of nodes (not necessarily links), flows are functions defined on the space of links L.

In transportation (and in other branches of engineering or science), it is more customary to think of flow as the amount of something passing a point or crossing a boundary per unit time. If the flow is changing with time, one ordinarily thinks of a flow $f_{ij}(t)$ as being an "instantaneous" rate:

$$f_{ij}(t) = [\text{amount crossing a point on the link } (i, j) \text{ in the time } t \text{ to } t + dt]/dt \tag{5.1}$$

or

$$f_{ij}(t) = \frac{d}{dt} [\text{cumulative amount to cross a point in a time } 0 \text{ to } t]. \tag{5.2}$$

These are only mathematical idealizations. Goods normally move in discrete amounts (numbers of cars), not like an infinitely divisible fluid. In practice one must choose the dt in (5.1) as a time that is not arbitrarily small but large enough so that the amount of goods in question has a meaningful interpretation. Or, if one uses (5.2), one must draw a graph of cumulative amount versus t, smooth the curve, and then evaluate the derivative (slope). There are also questions about reproducibility of observations (stochastic effects), but we will not be concerned with these more subtle problems of rigorous interpretation.

The more serious problems in large-scale transportation planning concern the practical question of whether one should study peak flows, average flows over a day, or perhaps yearly flows. Ideally, one should understand everything that is relevant to the design and use of transportation facilities, including the evolution of $f_{ij}(t)$ over time (hourly, weekly, seasonally) and the interrelations between the $f_{ij}(t)$ on different links. Given but a finite budget to make a transportation study, however, one must make compromises. If one treats a complex network, one cannot afford to investigate time dependence also.

Most transportation planning procedures are based on a sort of quasi-stationary flow pattern in which the $f_{ij}(t)$ are interpreted as either the total trips over a link (i, j) during 24 hours or the trips during the rush hour. These are not flows in the customary sense of hypothetical instantaneous flows but are time averages of such flows or, equivalently, just total trips (without necessarily the interpretation as a rate).

The actual flow is what interests planners, however. To go from 24-hour or rush-hour trips to a flow, planners usually use some arbitrary "conversion factor," saying that "from experience we find that x percent of the 24-hour trips occur between 8:00 A.M. and 8:30 A.M.," during which time the real flows are fairly steady.

Consider a single link, a single one-way road with one entrance and one exit as in figure 5.1. Let

$F^{(-)}(t)$ = total number of vehicles (or amount of goods) entering link from 0 to t,

$f^{(-)}(t) = dF^{(-)}(t)/dt$ rate of entering at time t,

Figure 5.1
Flow on a link.

$F^{(+)}(t)$ = total number of vehicles (or amount of goods)
leaving link from 0 to t,

$f^{(+)}(t) = dF^{(+)}(t)/dt$ rate of leaving at time t,

$N(t)$ = volume (number of cars, amount of goods) in the
link at time t.

If there are no other exits or entrances, and cars (goods) do not disintegrate, then there is a "conservation principle":

$$N(t) - N(0) = F^{(-)}(t) - F^{(+)}(t).$$

This conservation principle and its generalizations to networks form the single principle that is common to all aspects of traffic and transportation theory. The various branches of traffic theory are distinguished by the type of situation to which the principle is applied and the subsidiary conditions that are added to it. The "theory of network flows" deals with the generalization to networks of the special case in which $F^{(-)} = F^{(+)}$ or $N(t) = N(0)$ for all t. The theory of "traffic dynamics" or "stream characteristics" deals with the time-dependent properties of a network, the links of which are short sections of a single highway. This is supplemented by an "equation of state" relating $f(t)$ (the flow) to $N(t)$ (the density). "Queuing theory" deals with very simple types of networks in which there are subsidiary relations among $F^{(-)}$, $F^{(+)}$, and N usually involving stochastic properties.

At some time it would be desirable to develop practical methods for analyzing the propagation of surges over networks. At present, however, the study of the propagation of disturbances over networks has been confined to very simple networks (mainly a single highway), and the study of nontrivial networks has been confined to stationary flows.

The equilibrium condition $F^{(-)} = F^{(+)}$ (or its generalization to networks) could be justified on several grounds:

1. The link has no capacity—that is, $N(t) = 0$ for all t—or it has a negligible capacity relative to the particular problem being analyzed.

2. The flow pattern is stationary—that is, $N(t) = N > 0$ for all t—or it is approximately so over a relevant time period.

3. The evolution of the system is such that $N(t) = N(0)$ for particular values of t. For example, at $t = 24$ hours, the system may return to its condition at $t = 0$.

Of these arguments, only the third is really convincing, particularly if the times run from midnight to midnight; that is, if one is interested in 24-hour flows, one probably need not worry about the storage on the network. But by observing only 24-hour flows, one may be averaging out the very things that should be examined, such as rush hours.

It is easy to be drawn into thinking that the first two arguments may be true when, in fact, they are not. Looking at a mile-long section of freeway, for example, one might elect to observe flows for fifteen minutes or an hour. The flow during such a period may be 1,000 cars, whereas the fluctuation in the number of cars on the section may be less than 100. One might then conclude that the 100 is negligible compared with the 1,000, so that $F^{(-)} \approx F^{(+)}$. By extension, observations of flows $F^{(j)}$ at a sequence of points j along the freeway might be expected to show that $F^{(1)} \approx F^{(2)}$, $F^{(2)} \approx F^{(3)}, \ldots, F^{(n-1)} \approx F^{(n)}$. It does not necessarily follow, however, that $F^{(1)} \approx F^{(n)}$. Mathematically, the difficulty here is that the differences $F^{(j+1)} - F^{(j)}$ may all have the same sign. The quantity $F^{(n)} - F^{(1)}$ is the sum of all these small differences, but if n is large, the sum need not be small. In "physical" terms, although a mile of freeway has a negligible capacity for storage, this does not mean that ten miles of freeway have a negligible capacity compared with the flows, which would be similar for the one- or ten-mile sections.

There is a similar difficulty with the second argument. In a small region the flows near the peaks of the rush hours may remain fairly steady, yet the flow pattern will not be stationary as a whole because the peaks do not occur at the same time everywhere.

Although the following analysis is concerned almost exclusively with stationary flows, it should be applied to transportation planning very cautiously. There is a great danger that many of the implications planners draw from this theory are false. For example, it is neither necessary nor economically efficient to design a transportation system with sufficient capacity to accommodate the peak flows (which would be necessary if the demand were stationary). It is important that the peak flow lasts for only a finite time, that the system has storage capacity that allows queues to form at bottlenecks, and that traffic will get through even when demand temporarily exceeds capacity.

5.2
Forms of
Conservation
Equations

Unless we specify otherwise, we shall hereafter consider only systems in which we can neglect changes in the number of cars or goods in the network with time. We shall, in other words, imagine that the network is equivalent to one with no storage capacity. There may, however, be sources (origins) and sinks (destinations) external to the system.

The principle that no link in the system can store cars or goods can be translated into mathematical form in many ways:

1. First, the principle implies that on any directed link (i, j) the flows in and out are equal, so there is a unique number $f_{ij} \geq 0$ associated with that link. This is interpreted as the flow past any point along the link (if the link has a geometric interpretation as a line in space).

2. Suppose we consider only the flow going from an origin O to a destination D along a single route R from O to D. The natural extension of (1) would imply that there exists a unique number $f(R)$, the flow along R. Furthermore, this flow would exist on every link of R. Also, for trips traversing a cycle C, there is a unique number $f(C)$, the flow around C.

3. Suppose we have two or more routes or cycles, R_1, $R_2, \ldots$ and $C_1, C_2, \ldots$, such as in (2), each with its own flow $f(R_k)$ or $f(C_k)$. If there is a link (i, j) common to more than one R_k or C_k, then the flow on link (i, j) is

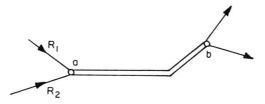

Figure 5.2
Two routes with a common
section.

$$f_{ij} = \sum_{k:(i,j)\in R_k} f(R_k) + \sum_{k:(i,j)\in C_k} f(C_k); \tag{5.3}$$

that is, the flow on any link (i, j) is the sum of the flows on all routes or cycles using the link (i, j).

The conservation principle looks so trivial here that one might wonder why the point needs to be labored. Indeed one must look at this rather carefully to see what is the principle and what is definition. The principle is essentially the statement in (2) that $f(R)$ is also the flow on every link of R. The equation in (3) is, in part, definition (that flows are additive), but it is compatible with the obvious requirement that if R_1 and R_2 have a common section of route, as in figure 5.2 from a to b, then the flow contribution from R_1 and R_2 is the same on all links of the subroute from a to b.

We have included the possibility of having a positive cycle flow. It is certainly mathematically acceptable that we have a flow circulating around a cycle, apparently going nowhere. For transportation flows one may be tempted to exclude this, but in examining something like a 24-hour flow there really is not much difference between a cycle flow, from home to work to home, and two trips, home to work and work to home.

Although this is a rather obvious formulation of the conservation principle, it is not the one that appears in most books, nor is it always the most convenient form. The difficulty arises because in any transportation network of reasonable size the number of routes R_k is astronomically large. Certainly one could not possibly analyze a network

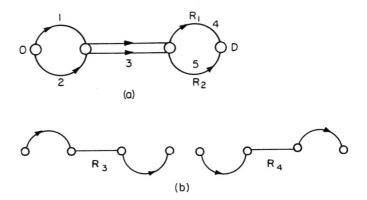

(a)

(b)

Figure 5.3
Link and route flows on a simple
network.

flow problem by explicitly comparing the consequences of
assigning traffic to routes in all possible ways. Actually, one
is not usually interested in the route flows anyway. One is
usually interested only in the total flows on links, and
probably only certain "critical" links. Because, as we shall
see, a large number of route flows usually give identical link
flows (for all links), it is perhaps advantageous to analyze the
link flows directly.

To illustrate the relation between link flows and route
flows, suppose we have two route flows between the same
origin and destination, and they have a link in common, as
in figure 5.3a. It obviously makes no difference in the flow
pattern if, after traversing the common link, two travelers
should interchange their subsequent routes to D. In terms of
route flows, however, this pattern is different from ones
involving routes like those in figure 5.3b.

Some things that are intuitively obvious from a picture
suddenly become quite cumbersome and abstract when
translated into formal mathematics or programmed for a
computer. Suppose we formalize the implied relations
between links flows and route flows. Let $f_1, f_2, \ldots, f_5$ be
the link flows on the five links, and let $f(R_1)$, $f(R_2)$, $f(R_3)$,
$f(R_4)$ be the flows on the four routes. Unfortunately in such
a small network one is likely to find that there are more links

than routes. While analyzing this however, one should imagine what will happen in more realistic situations for a complicated network with many possible routes through each link. For a network of n nodes, the number of links is likely to be some multiple of n (say $4n$), whereas the number of routes is likely to be on the order of c^n (for c comparable perhaps with 3). The number of routes (for $n = 100$ or more) is typically enormous (perhaps 10^{50}) compared with the number of links (perhaps 400).

From the figure one can see that

$$f_1 = f(R_1) + f(R_3), \tag{5.4}$$
$$f_2 = f(R_2) + f(R_4), \tag{5.5}$$
$$f_3 = f(R_1) + f(R_2) + f(R_3) + f(R_4), \tag{5.6}$$
$$f_4 = f(R_1) + f(R_4), \tag{5.7}$$
$$f_5 = f(R_2) + f(R_3). \tag{5.8}$$

This represents a system of simultaneous linear equations relating the f_j to the $f(R_k)$. Typically one would have many more $f(R_k)$ than f_j so that the $f(R_k)$ uniquely define the f_j but the converse is not true. Even in the present case in which we have five equations and only four $f(R_k)$, the f_j do not uniquely determine the $f(R_k)$. If we add equations (5.7) and (5.8) and subtract (5.6), or add equations (5.4) and (5.5) and subtract (5.6), we obtain

$$f_1 + f_2 = f_3, \tag{5.9}$$
$$f_4 + f_5 = f_3. \tag{5.10}$$

Because equations (5.4) to (5.8) can be interpreted as statements of the conservation principle on routes, equations (5.9) and (5.10) are a consequence of conservation. By themselves we recognize (5.9) and (5.10) as statements that the node has no storage capacity; what enters a node must come out. From the point of view of the theory of linear equations, we can consider (5.4) to (5.8) a system of linear equations that must be satisfied by the $f(R_k)$ for given f_j. We have five equations and four unknowns. But, in order that this system have a solution, it is necessary that the f_j satisfy two relations. If (5.9) and (5.10) are satisfied by the given f_j, only three of the five equations are independent. In effect we have only three equations for four unknowns. The solution is not unique. (This was immediately obvious before we wrote the equations (5.4) to (5.8).)

As a network becomes more complicated, the relations between link flows and route flows becomes more complicated. Unfortunately this relation is not just of academic interest. We cannot just pick one of these flow representations, say that it is the best one to use, and forget the others. We shall see later that certain traffic assignment procedures give route flows directly, whereas others give link flows. To compare one with the other, we must understand the relation between route and link flows. In particular, assignments in which each traveler chooses *his* optimal route usually lead to route flows, but assignments in which the total cost to society is minimized usually lead to a specification of link flows.

Typically the advantage of the route flow representation is that the conservation conditions are trivially satisfied, but the number of (even nonzero) route flows may be enormous. The advantage of the link flow representation is that it involves fewer variables; however, the conservation conditions sometimes cause awkward mathematical formulations.

5.3
One Origin,
One Destination

To develop a pattern of route flows such that a total flow q enters the network at an origin node O and leaves at a destination node D, let R_k be the kth route in some numbering of all routes $(n_1, n_2), (n_2, n_3) \ldots (n_{r-1}, n_r)$ with origin n_1 and destination n_r, and let $f(R_k)$ be the flow on R_k. The constraint that a flow q enters at O and leaves at D means that the $f(R_k)$ must satisfy

$$\sum_{\substack{k \\ \{R_k | n_1 = O, n_r = D\}}} f(R_k) = q, \tag{5.11}$$

$$f(R_k) = 0 \text{ if } n_1 \neq O \quad \text{or} \quad n_r \neq D, \tag{5.12}$$

$$f(C_k) \geq 0 \text{ for any cycle } C_k. \tag{5.13}$$

If link flows are obtained by superposition of route and cycle flows through equation (5.3), the link flows will necessarily satisfy a "node conservation principle." Because

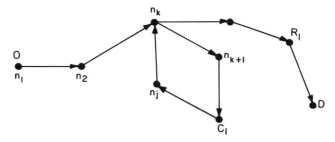

Figure 5.4
Identification of route and/or
cycle flows.

the flow into a node equals the flow out on each route except
at O and D, this is also true of the superposition. Thus if
f_{ij} is the flow on link (i, j),

$$\sum_{\substack{j: \\ (i,\,j) \in L}} f_{ij} - \sum_{\substack{k: \\ (k,\,i) \in L}} f_{ki} = \begin{cases} q \text{ if } i = \text{O}, \\ -q \text{ if } i = \text{D}, \\ 0 \text{ all other } i. \end{cases} \tag{5.14}$$

The first sum is the total flow out of node i; the second sum
is the total flow into node i.

Obviously any route flows define a unique set of link flows
satisfying (5.14). It is also obvious that if some link flows
satisfy (5.14), then, if there are some route flows that generate
it, they need not be unique; many route flows may give the
same link flows. We can, however, at least prove that for any
set of link flows, f_{ij}, satisfying (5.14), there exists a set of
route and cycle flows, $f(R_k)$ and $f(C_k)$, satisfying (5.3), (5.11),
(5.12), and (5.13). We first identify one possible route flow
from O to D or a cycle flow. Starting from O, there exists at
least one link (n_1, n_2), with $n_1 = \text{O}$, for which $f_{n_1 n_2} > 0$ by
virtue of (5.14). If there is more than one, any one will do.
Because there is a flow of at least $f_{n_1 n_2}$ entering n_2, there
must be a flow out of n_2, except possibly if $n_2 = \text{D}$. If $n_2 \neq \text{D}$,
choose a link (n_2, n_3) for which $f_{n_2 n_3} > 0$. Continue to
generate a sequence of links (n_j, n_{j+1}) for which $f_{n_j n_{j+1}} > 0$
until either $n_{j+1} = n_k$ for some k, $k < j$, or $n_{j+1} = \text{D}$; that is,
one forms a cycle $C_1 = (n_k, n_{k+1}), \ldots, (n_j, n_k)$ or a route
$(n_1, n_2), \ldots, (n_j, n_{j+1})$ from O to D, as in figure 5.4. Because

there are only a finite number of nodes and $n_1 \neq n_2, \ldots, n_j$, this must happen in a finite number of steps (less than the number of nodes n in the graph).

The smallest link flow on the cycle C_1 or the route R_1 is identified by

$$\min f_{n_k n_{k+1}}, \qquad (n_k, n_{k+1}) \in C_1 \quad \text{or} \quad R_1.$$

Because there are only a finite number of links in C_1 or R_1, this link flow is positive. From a cycle or route flow $f(C_1)$ or $f(R_1)$ having this value subtract the flow $f(C_1)$ or $f(R_1)$ from each link of C_1 or R_1 so as to form new link flows:

$$f_{ij}^{(1)} = \begin{cases} f_{ij} - f(C_1 \text{ or } R_1) & \text{if } (i,j) \in C_1 \quad \text{or} \quad R_1, \\ f_{ij} & \text{otherwise.} \end{cases}$$

This new set of link flows will also satisfy (5.14); however, q is reduced to $q - f(R_1)$ if a route is identified rather than a cycle. The $f_{ij}^{(1)}$ are all nonnegative,

$$0 \le f_{ij}^{(1)} \le f_{ij},$$

but there is at least one link $(i,j) \in C_1$ or R_1 for which $f_{ij} > 0$ but $f_{ij}^{(1)} = 0$.

Repeat this procedure for the link flows $f_{ij}^{(1)}$ so as to identify a cycle C_2 or route R_2 and a flow $f(C_2)$ or $f(R_2)$. Subtract this flow from the link flows $f_{ij}^{(1)}$ along C_2 or R_2 to obtain new link flows $f_{ij}^{(2)}$. By continuing this, a flow will be eliminated from at least one link in each iteration. In a finite number of steps (less than the number of links), the source flow q will be eliminated.

If there is still a link (i,j) with $f_{ij}^{(k)} > 0$ after q has been reduced to zero in the kth iteration, the procedure can be repeated with a slight modification. Start from node i and the link (i,j) and generate a sequence of links $(i,j), (j,k), \ldots$ with positive flows until one forms a cycle C_k (not necessarily returning to i). Subtract a flow $f(C_k)$ from this cycle and iterate until all link flows have been reduced to zero.

The set of cycle and route flows $f(C_k)$ or $f(R_k)$ generated, which necessarily satisfy (5.11), (5.12), (5,13), and the super-position of these, (5.3), give the flows f_{ij} satisfying (5.14). The actual route and cycle flows obtained will, in general, depend on the order in which links are selected to generate the routes R_k or cycles; they will not be unique.

The possibility of determining a set of $f(R_k)$ from a set of f_{ij} is conceptually quite important because many flow patterns are defined by the condition that they minimize some function (travel cost) of the $\{f_{ij}\}$, subject to the conservation principle. If one can determine an optimal set of f_{ij}, one must imagine the possibility of achieving this by assigning trips to specific routes.

5.4
Multiple Origin
and/or Destination The relation between link flows and route flows on a multiple O-D flow pattern is inherently more complicated than for a single O-D flow. For the latter each link flow pattern gives a family of possible route flows. For all practical purposes, the route flows are all equivalent, at least for flow patterns with no cycle flows. They differ only in that certain "identical" people traveling between the same origin and destination trade parts of their routes. In a multiple O-D flow, however, the family of route flows that gives the same link flow does not necessarily consist of "equivalent" flows.

Suppose, for example, that there was a route flow from an origin O_1 to a destination D_1 and another flow from O_2 to D_2, and the routes O_1–D_1 and O_2–D_2 crossed. The link flows would remain unchanged if, at the junction, some of the trips from O_1 to D_1 changed destinations with trips from O_2 to D_2 and thus generated route flows from O_1 to D_2 and O_2 to D_1.

Actually, two quite distinct types of situations exist depending on whether one considers these two flow patterns as equivalent. Consider, for example:

For given source and/or sink flows q_i at each node i, find the flow pattern $\{f_{ij}\}$ that conserves flow and minimizes the total journey distance (cost or whatever) of all trips.

For a given matrix q_{ij} of flows (trips) between origin i and destination j, find the flow pattern $\{f_{ij}\}$ that conserves flow and minimizes total distance.

The first problem is called the "transportation problem" in the mathematics literature. The second is what a transportation planner might call the "transportation problem." The first problem might arise in the following way. A manufacturer has several factories that produce identical goods at locations i, and they can ship amounts $q_i > 0$ from i. These goods are to be shipped to other nodes that "produce" $q_i < 0$ (actually consume $-q_i > 0$). The problem is to determine how much each producer will send to each consumer and along what route. Because the goods are identical and are not labeled as to producer, it makes no difference if goods from different origins trade destinations at the junction of two routes. The solution to the problem will be one in which an attempt is made to ship goods from a producer to the nearest consumers.

Unfortunately, in the transportation planner's problem, a person who starts a trip from node i is not willing to go to *any* destination. Throughout his trip he, in effect, carries a label designating where he wishes to go. When he crosses paths with someone else, he is not willing to trade destinations. The problem is equivalent to what in the mathematics literature is called a "multicommodity" flow. In the context of the manufacturers, suppose each factory produces a variety of different products in different amounts; for example, each factory produces a different product. Suppose each consumer also consumes specified amounts of each product; for example, the producer at i who makes a product i must ship an amount q_{ik} to consumer k. The question is: What routes should be used?

To represent a multiple O-D flow as a system of route flows involves only an obvious generalization of the single O-D representation. For each origin i and destination j with a flow q_{ij}, we can generalize equation (5.11) to

$$\sum_{\substack{k \\ \{R_k | n_1 = i, \, n_r = j\}}} f(R_k) = q_{ij}. \tag{5.15}$$

The total flow pattern is simply the superposition of route flows between each O-D pair i, j (plus perhaps some cycle flows).

The total link flows also satisfy a conservation equation that is the obvious generalization of (5.14), with the right-hand side replaced by the net source flow at node i to all destinations (sinks being counted negative); that is,

$$\sum_{\substack{j: \\ (i,\,j)\,\in\,L}} f_{ij} - \sum_{\substack{k: \\ (k,\,i)\,\in\,L}} f_{ki} = \sum_{j} q_{ij} - \sum_{j} q_{ji} \equiv q_i. \qquad (5.16)$$

The difficulty is that for a set of f_{kj} that satisfies (5.16) with given q_{ij}, there need not be a set of $f(R_k)$ satisfying (5.15). To overcome this, it is necessary to recognize that this is a multicommodity flow and each commodity itself satisfies a conservation equation.

The purpose in going from a route flow representation to a link flow representation was to reduce the number of parameters; there are many more $f(R_k)$ than f_{ij}. The route flows themselves satisfy a conservation principle, and if each route were identified as a "commodity," the multicommodity link flow representation would be the same as the route flow representation. If a multicommodity representation is introduced, one should try to use as few commodities as possible. Certainly if we identified each O-D pair as a separate commodity, the total flow would simply be the superposition of single O-D flows, each commodity having both route and link flow representations related to each other, as described in section 5.3. Whereas the number of routes is likely to be of order c^n for a network of n nodes, the number of commodities (O-D pairs) is of order n^2, and the number of link flows for each commodity is of order n. The total number of commodity link flows would be of order n^3.

Actually, it is not necessary to treat each O-D pair as a separate commodity; it suffices to identify commodities only by their origins (or only by their destinations). The number of flow variables need only be of order n^2. The reason for this is that everything that was said about the single O-D flow has an obvious generalization to a single origin– multiple destination or multiple origin–single destination flow; that is, the latter are, in effect, single commodity flows. If, in particular, one has a multiple set of origins with flows q_i and a single destination with a sink Σq_i, and one finds a set of link flows that satisfy the appropriate conservation equations at each node, then it is possible to find route flows

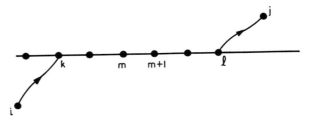

Figure 5.5
Trips entering and leaving a
single road.

with the correct total flows between each origin and the
destination that will generate the given f_{ij}. All the possible
patterns of route flows are essentially equivalent in the sense
that we do not care if trips with different origins but the same
destination trade paths at some point of route intersection to
the common destination. The only reason we must introduce
multicommodity flows to treat the multiple origin–multiple
destination flow is to prevent trips with different origins and
different destinations from trading paths.

**5.5
Flows on a
Long Road**

This rather abstract description of flow patterns on an
arbitrary graph does not explicitly recognize the fact that
most of the main transportation facilities of a metropolitan
region are actually long routes intersected by various
entrances and exits. In the representation of the network by a
graph, this long route is typically broken into a sequence of
links between successive entrance and exit points (nodes)
along its path. These links are treated as separate units like
any other links, disregarding the obvious geometric
continuity of the facility.

Although, in principle, it is possible that a trip from some
origin i to a destination j might choose a route that uses
many disjoint segments of this facility, most trips will use
only one continuous stretch of it, entering and leaving it
only once. If it is certain that every trip from i to j will enter
the facility at some point k and leave it at a point l, as
illustrated in figure 5.5, then one could consider the trips
from i to j as if each were three separate trips, from i to k,
k to l, and l to j. Correspondingly, the original O-D flow q_{ij}

could be replaced by three flows,

$$q_{ik}^{(i,\ j)} = q_{kl}^{(i,\ j)} = q_{lj}^{(i,\ j)} = q_{ij},$$

in which, for example, $q_{kl}^{(i,\ j)}$ represents the number of trips from k to l with primary origin i and ultimate destination j, and with k and l uniquely defined functions of i and j.

If there were originally flows q_{ik}, q_{kl}, q_{lj} between the nodes i, k, l, and j, then the $q_{ik}^{(i,\ j)}$, $q_{kl}^{(i,\ j)}$, $q_{lj}^{(i,\ j)}$ would be added to these to give a new O-D matrix, but the q_{ij} would be subiracted to give no trips from i to j. Actually the original O-D matrix q_{ij} is usually considered to be nonzero only if i and j are centroids. Consequently, nodes on our facility will not generally be primary origins or ultimate destinations; that is, q_{ik}, q_{kl}, and q_{lj} will all be zero.

If we knew where every trip that used our facility would enter and leave (that is, for each i and j, we knew a k and l, and a $q_{kl}^{(i,\ j)}$), then the flows along the facility would be determined entirely from an effective O-D table of entrances and exists from the facility:

$$q'_{kl} = \begin{cases} \text{total number of trips (per unit time) that enter} \\ \text{the facility at } k \text{ and leave at } l, \\ \displaystyle\sum_{i,\ j} q_{kl}^{(i,\ j)}. \end{cases} \qquad (5.17)$$

This summation over i, j is over all of the O-D pairs for which the trips from i to j enter and leave at the specified values of k and l.

Of course, one could determine the existing q'_{kl} directly from an O-D survey of trips entering or leaving the facility (a roadside survey) without measuring the complete q_{ij} (from home interviews). It might also be possible to make estimates of the q'_{kl} for some projected travel pattern without a detailed accounting of all the $q_{kl}^{(i,\ j)}$. If one were interested in only the future flow pattern along a single facility and had not made a complete O-D survey of the surrounding region, one obviously would need to find a cheaper way of estimating the q'_{kl} than from a projected q_{ij}.

If one knew the q'_{kl}, the evaluation of the flow on any link of the facility would be very simple. It might be convenient, however, to number the nodes of our facility consecutively so that all links were of the form $(m, m + 1)$. According to (5.3), the flow on the link $(m, m + 1)$ includes the flow on any route R containing the link $(m, m + 1)$. In terms of the present representation, however, the only routes using the link $(m, m + 1)$ are those between points k and l, namely $(k, k + 1), (k + 1, k + 2), \ldots, (m, m + 1), \ldots, (l - 1, l)$, with entrance on one side of the link and exit on the other side; that is, $k \leq m, l \geq m + 1$. Because these routes carry all the trips q'_{kl} between k and l, the link flow is

$$f_{m, m+1} = \sum_{l \geq m+1} \sum_{k \leq m} q'_{kl}. \tag{5.18}$$

Correspondingly, the "reverse flow" on the link $(m + 1, m)$ is

$$f_{m+1, m} = \sum_{l \leq m} \sum_{k \geq m+1} q'_{kl}. \tag{5.19}$$

It is often convenient to think of the number pairs k, l as points in a two-dimensional lattice, $[k, l]$. The region of summation in (5.18) and (5.19) can then be represented by those points in the shaded areas of figure 5.6a.

The flows (5.18), (5.19) can also be described in other ways. Suppose for each entrance node k, we were to specify a trip length distribution. If we let

$q^*_{k, x} = q'_{k, k+x} = $ number of trips (per unit time) that enter at k and travel a distance of x links,

then

$$f_{m, m+1} = \sum_{x > 0} \sum_{k = m - x + 1}^{m} q^*_{kx}, \tag{5.20}$$

$$f_{m+1, m} = \sum_{x < 0} \sum_{k = m+1}^{m - x} q^*_{kx}. \tag{5.21}$$

If we consider the index pair k, x as a point $[k, x]$ on a two-dimensional lattice, the regions of summation in (5.20) and (5.21) are as shown in figure 5.6b.

The reason for writing $f_{m, m+1}$ in the form (5.20) is that one often makes the postulate that the trip length distribution is

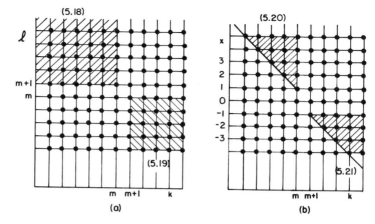

Figure 5.6
Regions of summation.

often makes the postulate that the trip length distribution is the same for all entrance nodes k (over some appropriate range of k); that is,

$$q_{kx}^* = q(x) \qquad \text{for all } k. \tag{5.22}$$

The summation over k in (5.20) can now be evaluated to give a single summation:

$$f_{m,m+1} = \sum_{x>0} x\, q(x), \tag{5.23}$$

$$f_{m+1,m} = \sum_{x<0} |x|\, q(x). \tag{5.24}$$

If we let

$$p_+(x) = \frac{q(x)}{q_+} \qquad \text{for } x > 0, \qquad q_+ = \sum_{x>0} q(x), \tag{5.25}$$

the $p_+(x)$ can be considered as a probability distribution of positive trip lengths, and

$$f_{m,m+1} = q_+ \sum_{x>0} x\, p_+(x) = q_+\, L \tag{5.26}$$

with

$$L = \sum_{x > 0} x \, p_+(x).$$

The L is an expectation (average) of the positive trip lengths along the facility. In this form, (5.26) conveniently demonstrates that the flow past any point or along a link depends not only on the number of trips made, q_+, but also the (average) length of the trips.

If the entrance and exit points are close together compared with the length of a typical trip, it may be convenient to idealize the discrete set of nodes and links by a continuum. We could let x represent any continuous labeling of points along the facility; for example, x could represent the number of nodes or fraction thereof, or it could represent the actual travel distance from some arbitrary origin.

We can now define a joint density $\rho'(x, y)$ so that

$$\rho'(x, y) \, dx \, dy = \text{number of trips (per unit time) that}$$
$$\text{enter the facility between } x \text{ and } x + dx$$
$$\text{and leave between } y \text{ and } y + dy. \quad (5.27)$$

The dx and dy cannot be "arbitrarily small" in the usual mathematical sense. They are small compared with a trip length but still must include at least several nodes, unless we "spread" the trips entering any node over the link between nodes. There is, of course, the implication here that the number of trips entering or leaving some (short) interval is (more or less) proportional to the length of the interval.

This $\rho'(x, y)$ is the continuum analog of the q'_{kl} in (5.17). The continuum analog of the q^*_{kx} in (5.20) is

$$\rho^*(x, l) \, dx \, dl = \text{number of trips (per unit time) entering}$$
$$\text{the facility between } x \text{ and } x + dx$$
$$\text{having a trip length between } l \text{ and}$$
$$l + dl. \quad (5.28)$$

In terms of these densities, we can define a positive direction flow past a point z as

$$f(z) = \int\limits_{\substack{x < z \\ y > z}} \int \rho'(x, y) \, dx \, dy \tag{5.29}$$

or

$$f(z) = \int_0^\infty \int_{z-l}^z \rho^*(x, l) \, dx \, dl, \tag{5.30}$$

the continuum analogs of (5.18) and (5.20), respectively.

If $\rho^*(x, l) = \rho(l)$ is independent of x, then (5.30) can be reduced to

$$f(z) = \int_0^\infty l \, \rho(l) \, dl \tag{5.31}$$

or

$$f(z) = \rho_+ \int_0^\infty l \, p_+(l) \, dl = \rho_+ \, L. \tag{5.32}$$

In (5.32),

$$\rho_+ = \int_0^\infty \rho(l) \, dl \tag{5.33}$$

is the number of trips (per unit time) entering per unit length traveling in the positive direction and

$$p_+(l) = \rho(l)/\rho_+ \tag{5.34}$$

is a probability density of positive trip lengths. The L is again the average trip length, but it is now measured in the units of the coordinate system x. The value of $\rho_+ \, L$ must, of course, be independent of the unit of length.

To illustrate the main point raised in section 5.4 regarding multicommodity flows, it suffices to note that a set of link flows $f_{m, m \pm 1}$, for all m, compatible with the node conservation equations

$$f_{m, m+1} - f_{m-1, m} + f_{m, m-1} - f_{m+1, m} = \sum_l q'_{ml} - \sum_k q'_{km},$$

for given values of the q_{kl} cannot necessarily be generated from a set of route flows corresponding to these q_{kl}. Even if we consider only trips in the positive direction, a set of $f_{m,m+1}$ satisfying

$$f_{m,m+1} - f_{m-1,m} = \sum_{l>m} q'_{ml} - \sum_{k<m} q'_{km} \tag{5.35}$$

cannot necessarily be generated from a set of route flows for the same set of q'_{kl}, $k < l$.

The issue here is quite clear, particularly in the case for which $q'_{k,k+x} = q^*_{kx} = q(x)$. If one only specifies the net (positive direction) flow into or out of a node or even specifies both of them separately—that is, $\Sigma q'_{ml}$ and $\Sigma q'_{km}$—one would have said virtually nothing about the mean trip lengths. A set of $f_{m,m+1}$ satisfying (5.35) could have come from a wide range of possible trip length distributions, not necessarily ones compatible with the given q_{kl}.

In the special case $q'_{k,k+1} = q(x)$,

$$\sum_{l>m} q'_{ml} = \sum_{x>0} q^*_{m,m+x} = \sum_{x>0} q(x) = q_+$$

and

$$\sum_{k<m} q'_{km} = \sum_{x>0} q^*_{m-x,m} = \sum_{x>0} q(x) = q_+.$$

Consequently (5.35) gives $f_{m,m+1} = f_{m-1,m}$; that is, the link flows are all the same. This is the only condition on the $f_{m,m+1}$; the actual value of the f_m could be as low as q_+ if each trip only went one link (or even zero if one allowed trips to go zero distance). In any case, the $f_{m,m+1}$ would not necessarily have the value specified by the given trip length distribution.

If one decomposes the flow into commodities by specifying a set of link flows associated with each entrance node—that is, a set of $f^{(k)}_{m,m+1}$, which is the link flow on link $(m, m+1)$ that entered at node k—then these link flows would uniquely determine the q'_{km} through

$$q'_{km} = f^{(k)}_{m,m-1} - f^{(k)}_{m+1,m}.$$

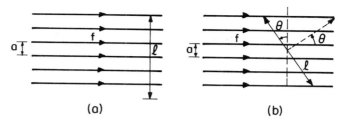

(a)   (b)

Figure 5.7
Evaluation of flux.

Thus there is a (unique) set of route flows associated with each set of commodity link flows.

For a network with a very large number of minor streets, one does not wish to describe the flow on every street. In the limit of an infinite number of streets on an arbitrarily fine grid, the flow on any single street may be negligible, but the total flow on a band of streets may be significant. We consider now the question of how one should represent flows on a continuum so as to satisfy conservation conditions.

Consider first a rectangular grid of arbitrarily small mesh size. Suppose that we have a single commodity and that the traffic is distributed over the network so that neighboring parallel streets carry nearly the same flow.

For the flow along a single family of parallel streets, the amount of flow across a line perpendicular to the streets is proportional to the length of the line. If each street carries a flow f, and the streets are spaced a distance a apart, as in figure 5.7a, then the flow across a line of length l is

Flow $\simeq (l/a)f = (f/a)l$.

The flow across a line per unit length is called a *flux*. In this example,

Flux $= F = f/a.$ \hfill (5.36)

A flux has all the properties associated with a vector. Note that if the line is not perpendicular to the street, as in figure 5.7b, then the flow across l is

Flow $= (f/a)l \cos \theta,$

where θ is the angle between the normal to l and the direction of the flow (or between l and the normal to the flow). The same result would be obtained if we took the flow or flux in the direction of the roads and resolved it into components parallel and perpendicular to the line l or its normal. (Whether we consider l itself a vector or represent it by a vector of length l in the direction of the normal to l is immaterial in two dimensions. It is customary, however, to represent it by a vector in the normal direction because most sciences in which these concepts are used deal with three-dimensional spaces and flows across surfaces rather than lines. In three dimensions a surface element is, of course, represented as a vector in the direction normal to the surface.)

Because the total flow across any line is the sum of the flows on various families of roads, it follows that the addition of fluxes can be considered to follow the same rules as for the usual addition of vectors, despite the fact that the resulting flux vector may not be in a direction in which one can actually travel. Suppose, for example, that we had a three-directional grid, as in figure 5.8, and there were fluxes $\bar{F}^{(1)}$, $\bar{F}^{(2)}$, and $\bar{F}^{(3)}$ along the three grid directions; let $\bar{F}^{(j)}$ be the flow crossing a unit length of line perpendicular to the direction of the jth family of roads. The overbar is used here to designate a vector quantity.

For any line of length l, as in figure 5.8, with a normal in the direction of the unit vector $\bar{n}$, the total flow across l is

$$l[\bar{F}^{(1)} \cdot \bar{n} + \bar{F}^{(2)} \cdot n + \bar{F}^{(3)} \cdot n],$$

in which the dot denotes the scalar product

$$\bar{F}^{(j)} \cdot \bar{n} = |F^{(j)}| \cos \theta_j$$

and θ_j is the angle between $\bar{F}^{(j)}$ and $\bar{n}$. If we let $\bar{F}$ be the usual vector sum of the $\bar{F}^{(j)}$,

$$\bar{F} = \bar{F}^{(1)} + \bar{F}^{(2)} + \bar{F}^{(3)}$$

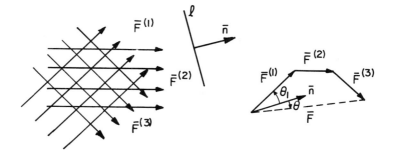

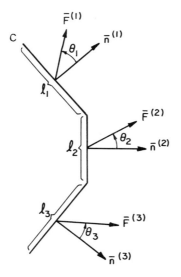

Figure 5.8
Addition of fluxes.

Figure 5.9
Flow across a curve.

as in figure 5.8, then the flow across l can also be written as

$$l[\bar{F}^{(1)} + \bar{F}^{(2)} + \bar{F}^{(3)}] \cdot \bar{n} = l\,\bar{F} \cdot \bar{n} = l\,|F|\,\cos\theta,$$

where θ is the angle between $\bar{F}$ and $\bar{n}$.

   If one has any piecewise linear curve C with segments of
lengths $l_1, l_2, \ldots$, and unit normals $\bar{n}^{(1)}, \bar{n}^{(2)}, \ldots$, as shown in
figure 5.9, the total flow across C is, by definition, the sum of

the flows across its component segments:

$$\text{Flow across } C = l_1 \, \bar{F}^{(1)} \cdot \bar{n}^{(1)} + l_2 \, \bar{F}^{(2)} \cdot \bar{n}^{(2)} + l_3 \, \bar{F}^{(3)} \cdot \bar{n}^{(3)} + \ldots$$

If C is any smoothly varying curve, it can be approximated by a piecewise linear curve of segments l_j and the limit taken as $l_j \to 0$. This leads to the definition of a line integral,

$$\text{Flow across } C = \int_C \bar{n}(s) \cdot \bar{F}(s) \, ds, \tag{5.37}$$

in which ds is an infinitesimal length along C, s labels a point on the curve C, and $\bar{F}(s)$ and $\bar{n}(s)$ are the flux and normal vectors at that point. If C happens to be a closed curve or "cordon line" enclosing a region A, then

$$\text{Total flow out of } A = \int_C \bar{n}(s) \cdot \bar{F}(s) \, ds.$$

The conservation equation must say that for a closed curve C enclosing a region A,

$$\int_C \bar{n} \cdot \bar{F} \, ds = \begin{cases} \text{rate of trips crossing } C, \\ (\text{rate of generation of trips in } A) \end{cases}$$
$$- (\text{rate of termination of trips in } A)$$
$$- (\text{rate of change of the total number of travelers, cars, or goods in } A).$$

If we have an equilibrium, no capacity or equivalent, then the last term will vanish (it is the generalization of the $N(t)$ in section 5.1); however, we shall keep it for now.

If it is reasonable to approximate the streets by a continuum, it may also be appropriate to assume that the trip origins and/or destinations are distributed over space with some two-dimensional density. Let

$\bar{x} = [x_1, x_2]$ = vector position in space;

$Q(\bar{x}, t) \, dx_1 \, dx_2$ = number of trips per unit time starting minus those ending in the area between x_1 and $x_1 + dx_1$, x_2 and $x_2 + dx_2$ at time t;

$$Q(\bar{x}, t) = \text{net density of trips per unit time starting}$$
$$\text{at } \bar{x} \text{ at time } t.$$

Then the difference between the rate of generation of trips in A and the rate of termination of trips in A is

$$\iint_A Q(\bar{x}, t) \, dx_1 \, dx_2.$$

If the system has a storage capacity that is distributed, there may be a two-dimensional density of travelers (goods). Let

$$\eta(\bar{x}, t) \, dx_1 \, dx_2 = \text{number of travelers in } dx_1 \, dx_2 \text{ at time } t \text{ and}$$
$$\text{position } \bar{x}.$$

Then

$$\text{Total travelers in } A \text{ at time } t = \iint_A \eta(\bar{x}, t) \, dx_1 \, dx_2.$$

If $\bar{F}$, Q, and η are time dependent, the rate of change of the number of travelers in A is $(\partial/\partial t) \int \int_A \eta(\bar{x}, t) \, dx_1 \, dx_2$. The conservation equation now states that for any region A with boundary C,

$$\int_C \bar{n} \cdot \bar{F}(\bar{x}, t) \, ds = \iint_A Q(\bar{x}, t) \, dx_1 \, dx_2 - \frac{\partial}{\partial t} \iint_A \eta(\bar{x}, t) \, dx_1 \, dx_2.$$

$$(5.38)$$

Equations of this type appear in many branches of physics and engineering to describe the conservation of mass, momentum, energy, etc. In most of these areas the analogs of $\bar{F}$, Q, and η can be defined very accurately and are usually smooth functions of $\bar{x}$, t. Here we are not only approximating the counts of discrete vehicles by a continuum but also approximating the discrete streets by a continuum. Although one cannot measure $\bar{F}$, Q, and η on a scale of distance comparable with the spacing between vehicles on a single street or the spacing between streets, they may be fairly well defined on a much coarser scale of distance. Furthermore, on such a coarse scale, it may be possible to approximate them by smooth functions of $\bar{x}$ and t. The smoothing should be done in such a way that the smoothed versions of $\bar{F}$, Q, and η also satisfy (5.38) for any region R that is large compared with the mesh size.

In physics the conservation equations are more often written in the form of differential equations than in an integral form analogous to (5.38). If $\bar{F}$ can be differentiated with respect to x_1, x_2, then the divergence theorem can be applied to the left-hand side of (5.38) giving

$$\int_C \bar{n} \cdot \bar{F}(\bar{x}, t) ds = \int\int_A \text{div } \bar{F}(\bar{x}, t) dx_1 dx_2,$$

in which

$$\text{div } \bar{F}(\bar{x}, t) \equiv \frac{\partial F_1(\bar{x}, t)}{\partial x_1} + \frac{\partial F_2(\bar{x}, t)}{\partial x_2},$$

and $F_1(\bar{x}, t)$, $F_2(\bar{x}, t)$ are the components of $\bar{F}(\bar{x}, t)$ in the directions of x_1 and x_2, respectively. Equation (5.38) can now be written as

$$\int\int_A [\text{div } \bar{F}(\bar{x}, t) - Q(\bar{x}, t) + \partial \eta(\bar{x}, t)/\partial t] dx_1 dx_2 = 0$$

for all A. If this integral is zero for all A, then the integrand must vanish:

$$\text{div } \bar{F}(\bar{x}, t) = Q(\bar{x}, t) - \partial \eta(\bar{x}, t)/\partial t. \tag{5.39}$$

This is the form commonly used in physics to describe such things as the conservation of mass or energy, but this two-dimensional form of the conservation equation is not commonly exploited in transportation.

The one-dimensional form of (5.39) would be

$$\frac{\partial f}{\partial x} + \frac{\partial \eta}{\partial t} = Q(x, t), \tag{5.40}$$

in which f is the one-dimensional flow, η the one-dimensional density of travelers, and $Q(x, t)$ is the net rate of generation of trips per unit length. This equation has been used extensively to describe the propagation of disturbances along a highway (see references 7, 8, and 9).

One advantage of writing a conservative equation in the form (5.38) or (5.39) is that one may find it convenient to use other special coordinates, such as polar coordinates, rather than $[x_1, x_2]$. Suppose, for example, that a city has a compact center and a flow pattern (nearly) rotationally symmetric about the center. If $C(R)$ is a circle of radius R and the flux at R, $F(R, t)$, is always in the radial direction, then the left-hand side of (5.38) can be evaluated as

$$\int_C \bar{n} \cdot \bar{F}(\bar{x}, t)ds = \int_0^{2\pi} F(R, t)Rd\theta = 2\pi RF(R, t).$$

If $Q(\bar{x}, t) = Q(r, t)$, and $\eta(\bar{x}, t) = \eta(r, t)$, for $x_1^2 + x_2^2 = r^2$ (they are rotationally symmetric), then the angular integrations on the right-hand side of (5.38) can be evaluated giving

$$RF(R, t) = \int_0^R rQ(r, t)dr - \frac{\partial}{\partial t}\int_0^R r\eta(r, t)dr.$$

If we differentiate this with respect to R, we have

$$\frac{\partial}{\partial R}[RF(R, t)] = [RQ(R, t)] - \left(\frac{\partial}{\partial t}\right)[R\eta(R, t)]. \tag{5.41}$$

This has the same form as (5.40), except the f, η, and Q of (5.40) are replaced by RF, $R\eta$, and RQ. Equation (5.41) could be used to describe the propagation of disturbances in polar coordinates much as (5.40) is used in one dimension.

For time-independent flows, the last term of (5.41) vanishes. If, for R larger than some radius R_0, $Q(R, t)$ also vanishes, then (5.41) would specify that

$$\frac{d}{dR}[RF(R)] = 0;$$

that is,

$$F(R) = \frac{\text{const}}{R}, \qquad R > R_0;$$

the flux decays with distance as $1/R$. This "spreading" of the flux is relevant to the analysis of a morning or evening rush hour for trips to or from the city center ($R < R_0$). Note that this result is independent of the detailed geometry of the road network and is a simple consequence of the fact that any flow crossing a circle of radius R_0 must also cross a circle of radius R.

As was true for discrete networks, the continuous version of the conservation equations are valid for the totality of all traffic or any component of the traffic that maintains its identity (that is, for any commodity). In most applications, however, it is necessary to consider separately those components of the flow having the same origin (or destination) and consider the total flow as the sum or integral of the component flows over all origins (or destinations). For either a discrete or continuous network, the conservation equations themselves do not recognize that equal but opposite flow on a two-way street is not equivalent to zero flow (unless the one-way flows are identified as different commodities).

**5.7
Continuity of
Fluxes**

One of the main problems in trying to analyze flows on a real transportation network is to find some realistic way of determining the flows on the major facilities (which may be congested) without worrying any more than necessary about the flows on the much larger number of minor roads. The use of fluxes to describe the traffic on minor roads has some appeal because it describes only a coarse property of the flows not explicitly dependent on the detailed geometry of the roads (yet it certainly describes it in more detail than simply replacing all minor roads by a few dummy links to centroids). Whether one can estimate these fluxes is, of course, another matter.

We described in 5.5 how, in principle, an effective O-D table could be generated along a single facility by decomposing a trip into a part along the facility and two access trips. We even considered a near continuum of entrance

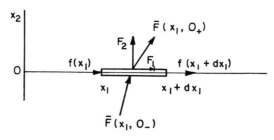

Figure 5.10
Continuity of flux.

and exit points along the facility that would imply a near continuum of access links; that is, a major road embedded in a near continuum of minor roads. To complete the discussion of conservation equations, we should also describe the interface between a major road carrying a flow and access roads described by fluxes.

Suppose we choose a coordinate system $[x_1, x_2]$ such that the major road is along the x_1-axis, coordinates $[x_1, 0]$, as shown in figure 5.10. If the road is two way, let $f(x_1)$ represent the net flow passing a point x_1, the flow in the positive direction minus that in the negative direction (if any). Suppose also that the flux vector $\bar{F}(x_1, x_2)$ is a continuous function of x_1, x_2, except that it may have a discontinuity along the line $x_2 = 0$ with values

$$\bar{F}(x_1, x_2) \to \bar{F}(x_1, 0_+) \qquad \text{for } x_2 \to 0, \quad x_2 > 0,$$
$$\bar{F}(x_1, x_2) \to \bar{F}(x_1, 0_-) \qquad \text{for } x_2 \to 0, \quad x_2 < 0.$$

Consider a rectangular slit of width dx_1 enclosing a section of the road, as in figure 5.10. If there is no storage on the road, the total flow crossing the boundary of the slit must be zero. In the limit in which the upper boundary of the slit approaches the x_1-axis from above, the flow across this boundary (out of the slit) approaches $F_2(x_1, 0_+)dx_1$. Similarly the flow across the lower boundary out of the slit has a limiting value of $-F_2(x_1, 0_-)dx_1$. The flows across the ends of the slit at x_1 and $x_1 + dx_1$ approach the values $-f(x_1)$ and $+f(x_1 + dx_1)$, respectively.

The total flow out of the slit is therefore

$$[F_2(x_1, 0_+) - F_2(x_1, 0)]dx_1 + f(x_1 + dx_1) - f_1(x_1) = 0.$$

If we divide this by dx_1 and let $dx_1 \to 0$, we obtain

$$\frac{df_1(x_1)}{dx_1} = F_2(x_1, 0_-) - F_2(x_1, 0_+). \tag{5.42}$$

If the flows are time dependent and there is storage space on the road with a density of $\eta(x_1, t)$ travelers, then (5.42) can be generalized to

$$\frac{\partial}{\partial x_1} f_1(x_1, t) + \frac{\partial}{\partial t} \eta(x_1, t) = F_2(x_1, 0_-) - F_2(x_1, 0_+), \tag{5.43}$$

which is essentially the same as (5.40).

One will notice that (5.42) or (5.43) involves only the component of the flux perpendicular to the road. Except at a major facility, the conservation equations will not permit a discontinuity of a flux component along any line perpendicular to that component. One may, however, have a discontinuity of a flux component along a line parallel with the flux component. Thus in figure 5.10 one may have any discontinuity in $F_1(x_1, x_2)$ along the line $x_2 = 0$.

One should again remember that these equations can be applied to the totality of all flows and fluxes or to any commodity flows. In most cases, one would apply them to separate commodities. Problem 1 illustrates how these concepts might be applied to a hypothetical city with many ring roads but only a few (major) radial roads.

Problems   **1**

A city contains n equally spaced ring roads at radius $j\,R/n$, $j = 1, 2, \ldots, n$, and four radial roads at angles $k\pi/2$, $k = 0, 1, 2, 3$. People reside only along the ring roads and generate trips at a constant density of ρ' trips per unit length of ring road. All trips end at the city center. If every trip is along the shortest route, find the flow (number of trips) passing each point on a ring road, coordinates $j\,R/n$, θ, and a radial r, $k\pi/2$.

For $n \gg 1$ (or the limit $n \to \infty$), determine the approximate flux vector at a point with polar coordinates r, $\theta(\theta \neq k\pi/2)$ and the approximate flow $f(r)$ at radius r along the radials. If there is a uniform density ρ trips generated per unit area in the circle of radius R, verify that the relation (5.42) is satisfied along the radial $\theta = 0$.

2

Suppose that in a two-dimensional plane the number of trips originating in an element of area $dx_1 dx_2$ is $\rho dx_1 dx_2$, independent of $[x_1, x_2]$. If trips from any origin have destinations uniformly distributed over a circle of radius L around the trip origin, how many trips cross a north-south (N-S) screen line per unit length from the west to the east side (the eastbound flux)?

3

On a square grid of roads, a freeway is built parallel to one of the grid directions. Trip origins are uniformly distributed in the plane with a density ρ trips per unit area per unit time. Each trip has a destination at a distance X_2 units perpendicular and X_1 units parallel to the direction of the freeway from its origin, $X_1, X_2 > 0$. A traveler is assigned to use the freeway if and only if he can do so with no increase in travel distance. The freeway has limited access with interchanges spaced D apart. What is the flow on the freeway as a function of X_1, X_2 and D?

Determine the flow on the freeway if X_1, X_2 have a distribution such that the number of trips per unit area with $x_1 < X_1 < x_1 + dx_1, x_2 < X_2 < x_2 + dx_2$ is

$$\rho L^{-2} dx_1 dx_2 \exp(-[x_1 + x_2]/L), \qquad \text{for } x_1, x_2 > 0.$$

4

On a fine square grid of two-way roads running N-S and E-W, is superimposes a high-speed road (two-way) running E-W and another running N-S, intersecting at $[0, 0]$. The number of trips originating in an element of area $dy_1 dy_2$ at $[y_1, y_2]$ destined for an element of area $dx_1 dx_2$ at $[x_1, x_2]$ is

$$\left(\frac{1}{4}\right)L^{-2}\rho\,dy_1\,dy_2\,dx_1\,dx_2\exp\left[-\frac{|x_1-y_1|}{L}-\frac{|x_2-y_2|}{L}\right],$$

in which ρ (the density of trip origins) and L (the mean trip component) are independent of y_1, y_2.

A trip may join a high-speed road at any point along the road. Each trip from $[y_1, y_2]$ to $[x_1, x_2]$ is assigned to the fastest route between $[y_1, y_2]$ and $[x_1, x_2]$, subject, however, to a restriction that the route must also be of minimum length, $|x_1 - y_1| + |x_2 - y_2|$.

How many trips pass through the intersection $[0, 0]$? Determine the flow $f(x_1)$ on the E-W fast road at $[x_1, 0]$ as a function of x_1.

References Any of the books on graphs listed in chapter 2 also discuss flows.

The use of continuum flows in two dimensions has seen little use, but see

1
Beckmann, M. J. "A Continuous Model of Transportation," *Econometrica* 20 (1952): 643–666.

2
Beckmann, M. J. "On the Optimal Location of Highway Networks," *Quantitative Geography.* Part I, edited by W. L. Garrison and I. F. Marble. Northwestern University Press, 1967.

3
Williams, H. C. W. L., and Ortuzar Salas, J. D. "Some Generalizations and Applications of the Velocity Field Concept: Trip Patterns in Idealized Cities," *Transportation Research* 10 (1976): 65–74.

The use of the term "flux" is borrowed from fluid dynamics and is not common in transportation. Miller and Holroyd use something analogous to a flux under the designation of "flow density."

4
Holroyd, E. M., and Miller, A. J. "Route Crossings in Urban Areas," *Proceedings of the Third Conference Australian Road Research Board* 3, part I (1966): 394–419.

5
Holroyd, E. M. "Routing Traffic in a Square Town to Minimize Route-crossings," *Beiträge zur Theorie des Verkehrsflusses*, Proceedings of the Fourth International Symposium on the Theory of Traffic Flow and Transportation (Karlsruhe, 1968).

A first attempt at a description of time-dependent flow on a city has been made in

6
Pearce, C. E. M. "Time Dependence in Commuter Traffic Models," *Transportation Science* 9 (1975): 289–307.

For a discussion of one-dimensional, time-dependent flow, see

7
Lighthill, M. J., and Whitham, G. B. "On Kinematic Waves II: A Theory of Traffic Flow on Long Crowded Roads," *Proceedings of the Royal Society* (London) 229A (1955): 317–345.

8
Edie, L. C. "Flow Theories," *Traffic Science*, edited by D. C. Gazis, pp. 1–108. New York: John Wiley & Sons, Inc., 1974.

9
Haberman, R. *Mathematical Models, Mechanical Vibrations, Population Dynamics and Traffic Flow*. Prentice-Hall, Inc., 1977.

6 TRAFFIC ASSIGNMENT

6.1
Introduction

Despite the fact that in a large network there is a large number of conservation equations to be satisfied, the dimensionality of the space of possible link flow patterns is very large, so large that the reduction of this dimensionality resulting from the conservation equations is not very helpful in cutting the problem down to manageable size. Flow patterns are generated by people deciding which of a large number of routes they will choose. The total flow is the superposition of such trips; and even though there may generally be many different decision schemes leading to equivalent total flow patterns, the range of possible flow patterns is very large.

In order to arrive at a uniquely defined flow patterns, one must make some rather strong assumptions regarding the manner in which real flow patterns are created (or should be created). This is a grand question of "model split." One usually uses the term "modal split" in reference to the choice between car, plane, train, etc., all of which could potentially be included as part of the network. The choice between different highways or transit routes, for example, involves the same type of questions of preferences.

Certainly travelers make a choice of route based, at least in part, on the metric or cost of the routes. If we were willing to postulate that each traveler had his own sense of values, thus his own private notion of a metric, then one could no doubt construct (perhaps by definition) a suitable set of values so that each traveler was certain to pick the route he considered best. Even if this were true, in principle, it would not help. It would be easier to observe what happens than try to infer what values each traveler has and why he does what he does.

We want to construct a logic for the collective behavior without getting involved in the details of individual motiva-

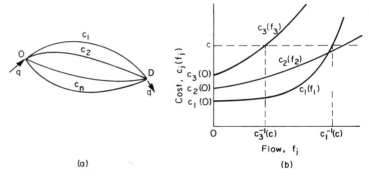

Figure 6.1
Travel costs and flows on n
parallel routes.

tion. We want to postulate that the total flow resulting from
individual and different motivations is in some sense similar
to what would result from a population of people with fairly
consistent motives or values.

Regardless of what people do, it is perhaps of some
interest to see what would happen if people were to behave
according to various hypothetical rules, preferably according
to some rules that lead to manageable mathematical formulation.

6.2
Single Origin–
Single Destination,
Cost Independent
of Flow

Consider the idealized network shown in figure 6.1a consisting of one origin, one destination, and two or more
possible routes between O and D. Let the cost, distance, or
whatever, between O and D be the same for each traveler,
$c_1, c_2, \ldots$ on routes 1, 2, $\ldots$, and suppose we wish to send
a total flow q from O to D.

The commonly discussed criteria for assigning the flow q
among the routes are:

1. Minimize the cost for each traveler.

2. Minimize the total cost for all travelers.

We consider these two assignment criteria first for the case
in which c_j is independent of the flow f_k on the kth route
for all k. Second, we consider the criteria for the case in
which c_j depends only on the flow for the jth route, f_j, but

not on f_k, $k \neq j$. If the criterion is (1), then everyone chooses the route (or routes) j such that

$$c_j = \min_k c_k. \tag{6.1}$$

This assignment gives

$$f_k = \begin{cases} q & \text{if } k = j, \\ 0 & \text{if } k \neq j, \end{cases} \tag{6.2}$$

provided j is unique; otherwise q can be distributed arbitrarily among the set of optimal routes. In the terminology of transportation planning, this is called the "all-or-nothing" assignment.

If the criterion is (2), then we wish to assign the flows f_k so as to minimize the total cost,

$$T = \sum_k f_k c_k, \tag{6.3}$$

subject to

$$\sum_k f_k = q \text{ (given)} \quad \text{and} \quad f_k \geq 0 \quad \text{for all } k. \tag{6.4}$$

Formally, this is a linear programming problem, minimizing a linear function of the f_k subject to linear equality or inequality constraints on the f_k. If the system of equalities and/or inequalities were sufficiently complicated, then one must use one of the well-known computer algorithms to find the f_k. Not all linear programming problems are difficult, however. The solution to this problem is the same as for (1) because, for any set of f_k satisfying (6.4),

$$T = \sum_k f_k c_k \geq \sum_k f_k c_j = c_j \sum_k f_k = c_j q.$$

Because this lower bound of T can be realized by the assignment (6.2), it must give the minimum of T.

6.3 Single Origin– Single Destination, Cost Dependent on Flow

Usually if one increases the flow on a network or facility in a transportation system, the travel time and, therefore, the effective cost per trip increases because of congestion. The cost of travel, however, must be interpreted as the cost for making a trip as compared with not making it on a facility that already exists. It does not include shared cost of the facility itself (except to the extent that tolls or other taxes are used for investment), which is likely to have the opposite type of behavior: The greater the capacity, the less the cost per unit of capacity, thus the greater the flow, the less the cost per trip. Neither does it include the cost of waiting for a transit vehicle if, in response to an increase in demand, a transit company were to add vehicles to the route. This would decrease the headway and the average waiting time and cause the cost per trip to decrease with flow.

Let us assume that the various routes of figure 6.1 are completely separated except at O and D so that the flow on one route does not physically interact with that on another route. The cost of travel on route j will be interpreted as a known (postulated or observed) function $c_j(f_j)$ of the flow f_j on route j, which is independent of the flows f_k, $k \neq j$, on other routes. We assume that $c_j(f_j)$ is a monotone increasing function of f_j for $f_j > 0$, as in figure 6.1b. If the route has a finite capacity, $c_j(f_j)$ will become arbitrarily large as f_j approaches the capacity and can be considered as infinite for f_j larger than the capacity.

We now reconsider the criteria of section 6.2 with this more general class of cost functions.

1. If every traveler chooses the route of minimum cost, we will again have, as in section 6.2, the obvious condition that if $c_m(f_m) > c_j(f_j)$ for some m and $f_m > 0$, then it is advantageous for some (or all) of the flow f_m to switch to route j. Because the $c_j(f_j)$ are assumed to be monotone increasing functions of f_j, any small transfer of flow from route m to j will reduce the cost on m and increase it on j. As long as $c_m(f_m)$ remains larger than $c_j(f_j)$ as the flow is shifted in small steps, the shift will continue until either all the flow has been removed from the expensive route ($f_m \to 0$) or the costs have been equalized. The only stable situation is

$$\begin{aligned} c_k(f_k) &= c && \text{on all routes with } f_k > 0, \\ c_k(0) &\geq c && \text{on all routes with } f_k = 0, \end{aligned} \qquad (6.5)$$

for c interpreted as the common cost of travel on all routes used. The value of c is dictated by the condition that $\Sigma_k f_k = q = $ given.

This does not determine the f_k "explicitly" as a function of q, but it is simple to evaluate the f_k either "analytically" or graphically. Because $c_k(f_k)$ is monotone increasing (and continuous) for $f_k > 0$, it has a unique inverse function; that is, for any number c, $c \geq c_k(0)$, there is a unique f_k such that $c_k(f_k) = c$:

$$f_k = c_k^{-1}(c) \qquad \text{for } c \geq c_k(0). \tag{6.6}$$

We can extend the definition of the inverse by defining

$$f_k = c_k^{-1}(c) \equiv 0 \qquad \text{for } c \leq c_k(0). \tag{6.7}$$

In effect, the function $c_k(f_k)$ includes a vertical segment at $f_k = 0$ from $c = -\infty$ to $c_k(0)$.

The curve $c_k^{-1}(c)$ is simply the curve for $c_k(f_k)$ with the axes interchanged or, equivalently, the same curve but with the vertical axis considered as the "independent variable" and the horizontal axis considered as the "dependent variable."

The value of c is now determined by the condition

$$\sum_k f_k = \sum_k c_k^{-1}(c) = q. \tag{6.8}$$

The function of c defined by

$$c^{-1}(c) \equiv \sum_k c_k^{-1}(c) \tag{6.9}$$

is a monotone nondecreasing function of c. It has an inverse that determines a $c(f)$ as a function of f. The solution of (6.8) for a given q is, therefore,

$$c = c(q), \tag{6.10}$$

and the flow on each route is given by (6.6) or (6.7).

This formal solution has a simple graphical construction, as illustrated in figure 6.2. Curves for $c_k(f)$ versus f for all k drawn on the graph would be interpreted as graphs of the inverse $c_k^{-1}(c)$.

141   Traffic Assignment

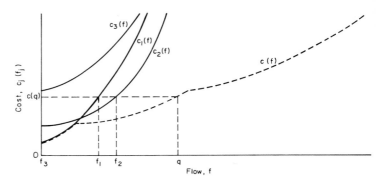

Figure 6.2
Partition of flows among n routes.

The addition of the inverse functions in (6.9) is essentially
the same as the usual addition of functions except that,
geometrically, for any value of c, we add all of the horizontal
distances $f_k = c_k^{-1}(c)$ at height c (instead of adding vertical
distances) and draw a graph of this horizontal sum $c^{-1}(c)$ as
a function of the vertical height c. This graph can also be
considered as the graph of $c(q)$ in (6.10).

To locate the f_k associated with any total flow q, a vertical line can be drawn to the curve $c(f)$ at flow q, then a
horizontal line at height $c(q)$. The flow values where this
horizontal line intersects the curves $c_k(f)$ are the f_k. The
construction automatically gives $f_k = 0$ if $c(a) < c_k(0)$.

In the special case in which there are only two routes
from O to D, a more commonly used graphical procedure
is to draw curves of $c_1(f)$ and $c_2(q - f)$ for a given value
of q, as in figure 6.3. If $c_1(f_1)$ and $c_2(f_2)$ include vertical
lines $c < c_1(0)$ at $f_1 = 0$ and $c < c_2(0)$ at $f_2 = 0$, these
curves will have a unique intersection where $c_1(f) =
c_2(q - f)$. This intersection determines $f_1 = f$ and
$f_2 = q - f$.

Although this construction can be used only for two
routes, it is somewhat simpler than that of figure 6.2 because
one need not perform the horizontal addition to construct
the curve $c(f)$. On the other hand, it does not show the
dependence of f_1 on q quite as conveniently nor does it

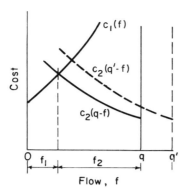

Figure 6.3
Partition of flows between two
routes.

show the curve $c(q)$ for the combined system, which is likely
to be useful for other purposes.

A single graphical construction in figure 6.3 determines
f_1 and f_2 only for one value of q. If q is changed to a q', the
graph of $c_2(f_2)$ simply can be drawn on a separate sheet of
transparent paper and turned upside down so that the flow
axis runs in the negative direction. This graph can be laid
over that for $c_1(f_1)$ so that the origin of f_2 is at $f_1 = q$. To
change q, one graph can be slid over the other.

A scheme of solution via graphs is very simple to perform
and illustrates all issues very clearly. It conveys information
much more efficiently (for most people) than an analytic
procedure that yields some formulas or a computer solution
that yields either a deck of cards or a table of numbers.
Unfortunately, the graphical procedures are not easily
generalized to describe flows on networks.

In comparison with the solution of section 6.2, for which
the functions $c_k(f)$ were represented by constants (horizontal lines), one might be tempted to conclude that it would
be only an "accident" if several c_k were equal; everyone
should take the fastest route. The more general solution in
this section, however, suggests that for sufficiently large q,
it will be typical for many routes to have the same travel
times.

One should be cautious about any idealization of a nonlinear optimization problem that leads to a linear programming problem. It is the nature of linear programming problems that the "optimal" solution be on a boundary. Frequently the idealization is the least accurate near a boundary. In the present case, the linearization is inaccurate if one tries to assign all the flow to one route that perhaps does not have the capacity to handle it.

2. The second type of assignment for flow q is to minimize

$$T = \sum_k f_k c_k(f_k),\tag{6.11}$$

with

$$\sum_k f_k = q \qquad f_k \geq 0.\tag{6.12}$$

In section 6.2, with $c_k(f_k) = c_k$, we saw that the two assignment principles gave the same flows f_k; this is no longer true if $c_k(f_k)$ increases with f_k.

Suppose we were to add a small flow ε to an existing flow pattern $\{f_k\}$ and add it all to the jth route (for some j); that is, $f_j \rightarrow f_j + \varepsilon$, f_k unchanged for $k \neq j$. The change in total cost T would be (for sufficiently small ε)

$$\varepsilon \frac{\partial T}{\partial f_j} = \varepsilon \frac{\partial}{\partial f_j}[f_j c_j(f_j)] = \varepsilon\left[c_j(f_j) + f_j \frac{dc_j(f_j)}{df_j}\right] = \varepsilon c_j^*(f_j),$$

with

$$c_j^*(f_j) \equiv c_j(f_j) + f_j dc_j(f_j)/df_j.\tag{6.13}$$

The first term, $c_j(f_j)$, can be interpreted as the cost of travel to a new traveler, as in (1); the cost for flow ε is ε times this. The second term is the cost that the new traveler imposes on others. By adding a flow ε, one has increased the cost to each of the original users by $\varepsilon dc_j(f_j)/df_j$. The quantity $c_j^*(f_j)$ is called the marginal cost (to society).

If the cost to society is to be kept as small as possible, society would ask the new traveler to go on the route having the smallest value of $c_j^*(f_j)$ rather than let him choose the route that is best for him, the one with the smallest $c_j(f_j)$.

To find the optimal assignment, notice that if we take a flow ε off route j and put it on route k, the change in T is

$$\varepsilon \left[c_k^*(f_k) - c_j^*(f_j) \right].$$

If both f_k and f_j are positive, then the cost will decrease if $c_k^*(f_k) < c_j^*(f_j)$. But if it were the other way around, we would switch a flow ε in the opposite direction. The only stable condition is

$$c_j^*(f_j) = c^* \qquad \text{for all } j, \quad f_j > 0; \tag{6.14}$$

that is, the marginal costs are equal on all used routes. The condition that some $f_j = 0$ is stable only if it does not pay to put a flow onto that route from any other:

$$c_j^*(0) = c_j(0) \geq c^* \qquad \text{for all } j, \quad f_j = 0. \tag{6.15}$$

The mathematical problem of evaluating the f_j as a function of q (that is, determining c^*) is exactly the same type as in (6.5) except that the $c_j(f_j)$ has been replaced by the marginal cost $c_j^*(f_j)$. Actually the uniqueness of the method in figure 6.2 depended on $c_j(f_j)$ being a monotone increasing function of f_j. That $c_j(f_j)$ is increasing does not guarantee that $c_j^*(f_j)$ is increasing, but it is reasonable to assume that the latter is true. The monotone behavior of $c_j(f_j)$ means that $dc(f_j)/df_j > 0$, which, in turn, implies that

$$c_j^*(f_j) \geq c_j(f_j) \qquad \text{for all } f_j.$$

For $c_j^*(f_j)$ to be an increasing function, it is necessary that

$$\frac{dc_j^*(f_j)}{df_j} = 2\frac{dc_j(f_j)}{df_j} + f_j\frac{d^2c_j(f_j)}{df_j^2} > 0.$$

Most cost functions, however, are convex—that is, $d^2c_j(f_j)/df_j^2 > 0$—so that $c_j^*(f_j)$ actually increases at least twice as rapidly as $c_j(f_j)$.

Typically as f_j approaches the capacity on route j, the marginal cost becomes very large. The travel time and, therefore, $c_j(f_j)$ may approach a finite limit, but $c_j^*(f_j)$

becomes infinite. If, for example, one route is a high-speed road and a second route is a low-speed road, then, according to criteria (1), everyone would use the high-speed road. If, however, the flow q is approaching the capacity of the first route, the marginal cost may be very high so that $c_1^*(f) > c_2^*(0)$ although $c_1(f) < c_2(0)$. This situation typically occurs between highways of quite different speeds (freeways versus city streets) because for low flows on the second route $c_2^*(f_2) \simeq c_2(f_2)$, whereas on the first route $c_1^*(f_1)$ could easily be twice as large as $c_1(f_1)$.

If, as we assumed, all travelers agreed on the relative merits of various routes—if they had a common set of costs $c_j(f_j)$—then they would undoubtedly distribute themselves approximately according to criteria (1), every man for himself, although it would be more desirable for society as a whole if they would distribute themselves according to (2). It is not obvious, of course, how the latter assignment would actually be achieved.

Some transportation economists have proposed that tolls be imposed in such a way that each traveler pays society (the government, the road authorities) the amount $f_j dc_j(f_j)/df_j$ to use route j. This is the cost that a "marginal user" of route j imposes on other travelers, and one might argue that a traveler should pay for the inconvenience he causes others. If one imposes tolls, however, every user of route j must pay the toll on route j (everyone is the marginal user). This toll, in effect, increases the cost of travel for everyone but is not considered a loss because society benefits from the revenue collected. This revenue would, presumably, be combined with other types of taxes to finance construction; it would be part of the "user tax," or it would be used to replace other forms of taxation.

To avoid any question as to what taxes this would replace, suppose that the toll on route j was the same amount less a rebate. The total revenue collected is distributed equally among all users of all routes so that some users pay a negative toll. Equivalently, the users of the better roads bribe their fellow travelers to use the poor route, and the toll collector becomes a nonprofit (zero expense) agent. The actual toll on route j is

$$f_j \frac{dc_j(f_j)}{df_j} - \delta,$$

with δ chosen so that the total revenue is zero:

$$\sum_k f_k^2 \frac{dc_k(f_k)}{df_k} - \sum_k f_k \delta = 0$$

or

$$\delta = \frac{1}{q} \sum_k f_k^2 \frac{dc_k(f_k)}{df_k}.$$

The rebate δ is a function of all the f_k so that the new cost per trip on route j, $c_j^*(f_j) - \delta$, no longer has the simple mathematical form in which it is a function only of f_j. Regardless of this, however, an equilibrium assignment according to criteria (1) must still require that the costs be equal on all routes used. But if the $c_j^*(f_j) - \delta$ are equal, so are the $c_j^*(f_j)$.

Despite the fact that the costs to individual travelers are changed by the toll, the δ has been chosen in a way that all new terms in T cancel:

$$\sum_j f_j [c_j^*(f_j) - \delta] = \sum_j f_j c_j(f_j) = T.$$

Thus the minimization of this with respect to the f_j yields the same assignment as before, but now the two assignments yield the same flows.

There are not many places where one could collect tolls, and, even if one could, the cost of the toll collection should be included in T. Aside from this, however, there are more serious social questions of subsidies for low income people, taxation according to ability to pay, and so forth. For the "optimal" distribution, someone must be assigned to less desirable routes. If people traveled the routes from O to D every day, then an alternative means of achieving the optimal distribution would be for people to take turns on good routes. The average cost over many days would be the same for everyone and less than if everyone chose the good routes. This assumes that people do not put value on uniformity of travel route from day to day (and that this could be implemented). The objection to taxation is that it implies a money equivalent for time. People who put a high value on their time will pay the toll and use the fast road. Low income people will, in effect, be assigned to the slow road. Whether this is proper is an interesting subject for debate, but it is not considered a part of the "theory of network flows." Of course, if people do not agree on the relative value of time and money, they will not agree on travel costs $c_j(f_j)$ either.

If we compare the mathematical formulation of the assignment criteria (1) and (2), we see that the minimization of an overall objective function in (2) led to the same type of equilibrium equations as in (1). This would seem to imply that there must be some hypothetical objective function whose minimization yields the same assignment as in (1). One can easily verify that the appropriate function is

$$T^* = \sum_k \int_0^{f_k} c_k(y)dy. \tag{6.16}$$

If we minimize this with respect to the f_k, instead of (6.11) subject to the conditions (6.12), we will obtain the equilibrium conditions (6.5) instead of (6.14). This follows immediately from the fact that the derivative of (6.16) with respect to f_j is $c_j(f_j)$, whereas the derivative of (6.11) with respect to f_j is $c_j^*(f_j)$.

There is no reason why T^* should have any economic significance, and, indeed, it does not appear to have any. The minimization of T^* is a consequence of the traveler's choice of routes, not a motivation for it. The individual travelers are thinking only of themselves, not of any global objective. One can, however, manufacture an artificial interpretation of T^*.

Suppose that, for any final choice of the f_k, we load the network incrementally. For each new increment of flow dy that we add to link k when there is already a flow y on the link, we evaluate the cost that the new traveler pays, $c_k(y)dy$,

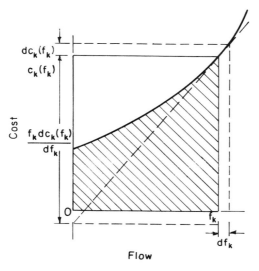

Figure 6.4
Interpretation of costs.

but disregard the increased cost that the new traveler imposes on the flow y already assigned to the link. The total cost paid by all travelers on link k is the sum of all these incremental costs; that is, $\int_0^{f_k} c_k(y)dy$. The total cost on all links is T^*. Because the cost does not include the cost that a new increment of flow imposes on the existing flows, the minimization of T^* is obviously achieved if each traveler is assigned to the cheapest route at each stage. Although certain computer assignment programs load networks incrementally, this is only a computational convenience. The interpretation of T^* is artificial because the basic premise of the theory is that all travelers are equal and all components of f_j are on the jth route simultaneously.

Figure 6.4 shows some geometrical interpretations of various quantities. For any graph of $c_k(f)$, the term of T^*, $\int_0^{f_k} c_k(y)dy$, is the shaded area under curve $c_k(f)$ from 0 to f_k. The corresponding term of T, the total cost $f_k c_k(f_k)$ on the kth route, is the area of a rectangle of base f_k and height $c_k(f_k)$ enclosing the shaded area. For any incremental change in flow df_k, the area of the vertical strip of base df_k and height $c_k(f_k)$ represents the cost paid by the increment of flow df_k. The area of the horizontal strip of height $df_k(dc_k(f_k)/df_k)$ and width f_k is the cost that the flow df_k imposes on others. The term $f_k dc_k(f_k)/df_k$ of (6.13) can also be determined by drawing a tangent line to the curve $c_k(f_k)$ at flow f_k, the broken line of figure 6.4 with the slope $dc_k(f_k)/df_k$. If this is extended back to flow 0, then the triangle of base f_k will have height $f_k dc_k(f_k)/df_k$.

In the assignment scheme (2), we started with an objective function, derived some equations that must be satisfied in order that the flows minimize this objective, and then solved the equations. This is the "classic" approach to optimization problems given in most mathematics books but one that in more complex situations cannot always be carried out (one may not be able to solve the equations derived from the optimization). For the assignment scheme (1), we started with a statement of principle that could be translated directly into a set of equations for the f_k (which we could solve), but then deduced an objective function T^* for which these f_k would give the minimum. It may seem as if we are artificially creating an optimization problem for which we already know the solution. In more complex situations, however, this will not be the case.

To minimize a function of one or more variables, one method is to set derivatives equal to zero (if the function has derivatives). This is what anyone who has studied elementary calculus is likely to do. Another method is simply to evaluate the function at many points and systematically find points that give smaller and smaller values to the function. If the function is "smooth," one might plot its values at many points on a graph, interpolate so as to draw a curve, and pick the lowest point on the graph. Equations from the first method may not be solved exactly, and the accuracy of approximate solutions is not obvious. In addition, the vanishing derivative, depending on the nature of the function, is neither a necessary nor sufficient condition for a minimum. The second method at least gives a sense that one approximate solution is "better" than another; it gives a smaller value for the function.

From a practical point of view, it is often very advantageous to know that the solution of a set of equations is equivalent to the minimization of some (even artificial) objective function. To obtain approximate solutions of the equations, it may be desirable to reverse the textbook procedure, to go from the equations to the objective function and find the approximate minimum of the objective function directly. This also may be advantageous from a theoretical point of view. It may not be obvious that the set of equations has a solution; yet it is obvious that a function has a minimum.

**6.4
Multiple Origin and/or Destinations, Costs Independent of Flows**

The previous section dealt with a trivial network consisting of one origin and one destination joined by a finite number of routes with no links in common. We determined how a specified flow q between O and D would be distributed among the routes according to certain proposed assignment schemes.

Suppose that we have a more complex network, an idealization of a real or proposed network, containing many origins and destinations. In following the transportation planning procedures, we have somehow determined an O-D matrix, q_{ij} trips from origin i to destination j (either for the present or the future). For various assignment schemes generalizing those of the previous section, we wish to determine the flows that would exist on various links of the network and the total cost of travel.

Imagine the network represented as a graph with a set of nodes and links. For any route R between an origin and a destination, there exists (under the prevailing conditions) a certain cost of travel $c(R)$. All users of this route perceive the cost in the same way and identify it as the sum of costs along the links of R,

$$c(R) = c_{n_1 n_2} + c_{n_2 n_3} + \ldots + c_{n_{r-1} n_r};$$

(6.17)

they see it as a "distance" in the sense of (3.7). Furthermore, travelers along any other route that traverses a link (k, l) see the same cost c_{kl} for this link.

The total flow f_{kl} on link (k, l) is the sum of all route flows using the link, and the total cost is the sum of all trip costs. The link travel cost c_{kl} does not depend on the flow f_{kl} along the link (k, l) or the flows on any other links. It is considered to be given from measurements of travel speed or whatever.

We will be primarily concerned with two idealized assignment schemes:

1. Every trip chooses the route of minimum cost.

2. Trips are assigned to routes so as to minimize the total cost of travel.

1. Because the c_{kl} are assumed to be known here, the shortest path algorithms discussed in chapter 3 can be used to determine both the cost and the cheapest route from every origin to every destination. The cheapest route from an origin O to any node k is actually identified only indirectly. The final labeling of node k gives only the cost (distance) from O and the previous node n_k from which k was reached by a cheapest path. However, from this one can trace the route by selecting the previous node to n_k, the previous node from that, and so forth, until one reaches O. In anticipation that the cheapest route will often not be unique, one can modify the algorithm so as to list all previous nodes that gave the least cost. From this, one can retrace all cheapest routes from O to k.

We saw in chapter 3 that the shortest routes from all origins to all destinations can be computed quite rapidly even for very large networks ($10^3 - 10^4$ nodes). The next step is to assign the flow q_{ij} to the cheapest route from i to j (or split it among the cheapest routes if it is not unique), and evaluate the total flow on each link. This can also be done very rapidly (in a computation time comparable with that for evaluating shortest routes). One method is to start at some destination node k (preferably an end of a tree from node i) and add the flow q_{ik} to the link flow on (n_k, k). To trace the route from k back to i, one may in the next step add the flow q_{ik} to that destined for n_k, q_{in_k}, and then add both to the flow on the link (n_{n_k}, n_k), the link preceding n_k (from i). As one retraces the route from k back to i through some node l, one can add to the link (n_l, l) the cumulative flow to all nodes between l and k (from i).

2. The minimization of the total cost of travel can be formulated in several ways. First one could write the total cost as a double sum. For each route R_k; (n_1, n_2) (n_2, n_3), ..., (n_{r-1}, n_r) from n_1 to n_r passing through a link (l, m), there is a cost $f(R_k)c_{lm}$ for $f(R_k)$ to traverse the link (l, m). The total cost of all trips is obtained by summing this over all links in R_k and then over all routes R_k:

$$T = \sum_{R_k} \sum_{(l, m) \in R_k} c_{lm} f(R_k). \tag{6.18}$$

This is minimized with respect to the $f(R_k)$, subject to the conservation conditions and nonnegative constraints:

$$\sum_{R_k : n_1 = i, \, n_r = j} f(R_k) = q_{ij}, \qquad f(R_k) \geq 0. \tag{6.19}$$

One can also collect the coefficient of $f(R_k)$ in (6.18) and write them formally as a single sum,

$$T = \sum_{R_k} c(R_k) f(R_k), \tag{6.20}$$

in which the $c(R_k)$ can be considered as "known" from (6.17), but the $f(R_k)$ are the "unknowns."

One can also interchange the order of summation in (6.18) and collect the coefficient of c_{lm} so as to obtain

$$T = \sum_{(l, m) \in L} c_{lm} f_{lm}. \tag{6.21}$$

This can now be minimized with respect to the link flows f_{lm}, which must satisfy conservation equations that guarantee that the link flows can be generated from route flows satisfying (6.19). As we saw in chapter 5, it is still necessary to represent the f_{km} as a sum of commodity flows in (6.21) and impose node conservation conditions on each commodity at each node. The commodities could label either the origins or the destinations (or both).

In chapter 5 we admitted the possibility that there were

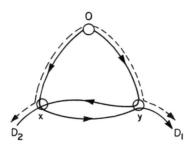

Figure 6.5
Rerouting of traffic.

cycle flows, and we showed that, for any commodity link flows satisfying the node conservation equations, there was at least one representation of this flow in terms of route flows plus cycle flows. Here we have discarded the cycle flows. We could have added some cycle flows to (6.18) or admitted the possibility that the travel path from i to j contained a cycle; but because the costs c_{lm} are all positive, the cost of traversing any cycle is positive. Consequently, the minimization of T with respect to the route plus cycle flows would automatically guarantee that all cycle flows are zero (and that no trip was assigned a path that included a cycle).

For a network of any reasonable size, there is an enormous number of routes; it may seem that the minimization of (6.18) would be a hopeless task; but, for the minimum cost assignment, most of the $f(R_k)$ are zero. Suppose, for example, as in figure 6.5, that there is a route flow along a route O to D_1 passing through x and y (in that order) and a route flow along a route from O to D_2 passing through y and x. If we take some flow off $Oyx D_2$ and send it via $Ox D_2$ and take an equal amount of flow off $OxyD_1$ and send it via OyD_1, the flows and costs on Ox, $x D_2$, Oy, and yD_1 will be unchanged, but the flows on xy and yx will be reduced. The new route flows and link flow patterns will satisfy the same conservation equations as the old flows, but the total cost has been reduced.

More generally, if there is a positive link flow of the same commodity (same origin or same destination) on all links of any cycle C, one can subtract from the link flow pattern a cycle flow equal to the smallest link flow on C (of that commodity). The resulting link flows will still satisfy the original conservation equations; consequently there will be at least one route flow assignment satisfying (6.19). The removal of this cycle flow reduces the total cost of travel. The minimum cost flow pattern will, therefore, have no cycle carrying a positive flow of the same commodity on all links.

This lack of cycle flows for each commodity in a network is true even if the link travel costs c_{lm} depend on the flows on all of the links, provided only that the total cost is an increasing function of all link flows (the elimination of any flow will reduce the cost). If the costs c_{lm} are independent of the flow, as in section 6.2, the assignment according to scheme (1) will also minimize T because, if any trip were not assigned to the least cost route, a shift to this route would decrease the cost of this trip but not change the cost of any other trip. As one can see from (6.20), this would reduce the total cost.

The simplest procedure for minimizing T is, therefore, to find the cheapest paths, as in (1), and, in effect, represent T as the sum of route flows. Despite the fact that there is an enormous number of possible routes and unknown $f(R_k)$, the shortest-path algorithm is extremely efficient in sorting out only a relatively small number of routes.

Although these assignment schemes can be worked very rapidly, they often lead to flows that are incompatible with the postulate that the c_{lm} are independent of the flows. The all-or-nothing assignment may send an impossible flow along certain routes, perhaps a few thousand cars per hour along an alley. It typically assigns zero flow to many links that are, in reality, used.

The more accurately one tries to specify the c_{lm}, the more likely it is that the cheapest route between i and j is unique; consequently, the cheapest route is assigned all of the flow q_{ij}. If one defines the c_{lm} to only two decimal places, then many routes might be equally good. One could then divide the flow q_{ij}, in some arbitrary way, among the equally good routes and perhaps avoid overloading any link. It is also true that the more realistic one tries to make the graph by inclusion of minor streets, the more likely the formal procedures will assign traffic to links that cannot handle it.

**6.5
Flow-Dependent
Travel Costs**

If the all-or-nothing assignment places too much flow on certain links, the assumption of a flow-independent cost must be modified. We corrected this for the simple network of section 6.3 by replacing the cost c_j on route j by a function $c_j(f_j)$. It was assumed that the flows on different routes did not physically interact, so it was reasonable to postulate that the cost on the jth route was independent of the flow on any other route. If there were only a few routes, one could also imagine doing an experiment to measure the $c_j(f_j)$. At various times of the day when there are different flows f_j, one could measure the travel time by way of route j, the fuel consumption, and so forth, and convert these into a proposed cost. One would perhaps also observe the actual distribution of flows among the routes and see how well the theory based on identical preferences of people checks with the observations (it is not likely to check very well).

Generalizing this to a complex network presents some problems in both the definition of costs and the mathematical solution of the assignment problem given the costs.

Although transportation studies might include some observations of travel time, for the most part the travel times or costs are inferred from a description of the facilities and some empirical formulas. For arterial highways or freeways, for example, if one specifies the lane width and number of lanes, one can evaluate the capacity from formulas and/or graphs in the Highway Capacity Manual or other traffic engineering handbooks. These books also give formulas for the travel speed as a function of the ratio of flow (volume) to capacity. They also describe capacities and travel times for highway intersections, freeway merges, synchronized signal systems, and so forth, more things than one would care to include explicitly in a gross description of the network.

In approximating a real transportation network by a graph of nodes and links, many streets are discarded and groups of roads are replaced by a single link. The idealized network does not include a detailed description of signal settings, merge geometrics, etc. Except possibly for certain high volume facilities (transit lines or freeways), nearly all links in the idealized network represent effective equivalents of groups of streets. These effective links have a capacity similar to the combined capacities of the real links that they represent and a cost of travel similar to that of any of the real links.

The final result of this idealization is that each link (l, m) of the network is given a cost function $c_{lm}(f_{lm})$ that depends on the flow f_{lm} on the link (l, m) but not on the flows on other links. This cost function is assumed to be a monotone increasing (and usually convex) function of the flow and, in effect, infinite for f_{lm} exceeding the assumed capacity of this link.

It is not clear to what extent this type of idealization gives an approximate description of the real network. It obviously gives a rather crude approximation of "local trips" but includes some important large-scale aspects. There is still some question, however, as to whether it properly describes even the dominant coarse features of real traffic.

One difficulty arises because there is considerable interaction between flows on neighboring links. Suppose, for example, that groups of parallel N-S and E-W streets of the real network are replaced by a single N-S and a single E-W road on the idealized network; that is, a four-way "effective" intersection. In the real network there is considerable interaction between N-S and E-W travel and turning movements at individual intersections. This single effective intersection should describe, approximately, the travel times, capacities, and so forth, of the real network.

It would probably be more realistic to say that the travel cost of the N-S or E-W directions depends on the *sum* of the N-S and E-W flows (or fluxes) than to say that the N-S costs depends on only the N-S flow and the E-W costs depend only on the E-W flow. If most of the trip time is spent at intersections that must accommodate all flows, it might be appropriate to define some effective density of trips in two dimensions and say that the travel cost per unit distance of travel is a function of this density.

A possible consequence of the assumption that c_{lm} is a

function of only f_{lm} is that the theory is likely to under-estimate the difficulty traffic has moving opposite to or across the predominant direction of traffic flow (as in a morning or evening peak) and is, therefore, likely to route more of this traffic through congested areas than it should.

Another difficulty with the theory is that the cost functions are expected to include some effective delays due to queues. Again, the postulate that the cost c_{lm} depends only on f_{lm} is not always satisfactory because a queue generated on one link may back up onto other links.

Suppose a uniform section of highway has a bottleneck; for example, a freeway with a merge point or an arterial highway interrupted by a traffic intersection. The bottleneck might be located at a node of the idealized network or it might be a part of a link. In the mathematical representation, travel costs are associated with links rather than nodes and are identified primarily as travel time. Consequently we would like to describe the travel time from one point, identified as a node upstream from the bottleneck, to another point, identified as a node at or downstream from the bottleneck, and show how this travel time or cost depends on the flow along this link (highway section).

Traffic engineering books contain many graphs of velocity versus flow for uniform sections of highways. These graphs are always two-valued functions of the flow, similar to the solid line curve in figure 6.6. The travel cost per unit distance is essentially 1/(velocity), travel time per unit distance, and is also a two-valued function of the flow, as shown by the broken line curve. The lower portion of the graph, $c_-(f)$, is the "free flow" state where the cost is a monotone increasing function of f. The upper portion, $c_+(f)$, is in the "congested" state, which occurs only in a queue upstream from some bottleneck.

If there is a bottleneck of capacity f_0, the flow past the bottleneck cannot exceed f_0. The flow approaching the bottleneck can exceed f_0 temporarily, but any excess traffic back up and causes a queue. The queue continues to grow until the travel time through the queue becomes so large that some travelers divert to other routes. When the queues propagate past some point along the highway, the velocity

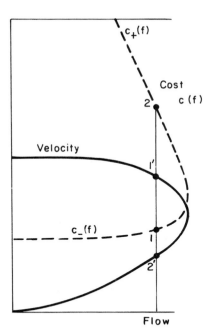

Figure 6.6
Velocity and travel cost on a
highway.

drops suddenly from some point near point 1′ of figure 6.6
to 2′, and the cost per unit distance of travel jumps from
1 to 2.

Suppose we have a link of length L_0 containing such a
bottleneck, and we let $c_+(f)$, $c_-(f)$ represent the cost per
unit length of travel over the congested and uncogested
parts of the link. At the time when the queue has propagated
a distance L upstream from the bottleneck, L less than the
distance to the next upstream node, the travel cost over this
link is

$$(L_0 - L)c_-(f_0) + Lc_+(f_0),$$

an increasing function of L. If there is an alternate route
bypassing this link with a travel cost between the values
$L_0c_-(f_0)$ and $L_0c_+(f_0)$, the queue length L should stabilize
at whatever value will make the cost along L_0 equal to that

on the alternate route. In effect, the travel cost function for the highway link is $c_-(f)$, for $f < f_0$, plus a vertical line from 1 to 2 in figure 6.6.

Actually the bottleneck acts somewhat like a stochastic service facility. Queueing delays occur, due to fluctuations, even for flows less than f_0. The average cost (averaged over the fluctuation in arrivals and service) should be a smooth function of f, which becomes very steep as f approaches f_0, at least for costs less than $c_+(f_0)$. Various empirical formulas used in the traffic assignment algorithms are intended to represent this average cost. (See, for example, reference 4.)

If the cost of travel by way of alternate routes were such that travelers would still prefer the highway link even if the cost were greater than $L_0 c_+(f_0)$, the queue on this link would propagate back to the upstream node and beyond. The queues would continue to grow on some or all links providing input to the bottleneck link until the excess demand for the bottleneck were diverted some place. For trips passing the bottleneck, the effective cost of travel would behave more or less as if the cost function $c(f)$ for the bottleneck link were infinite for $f \to f_0$. However, because, the queues would form on links that may also carry flow not destined for the bottleneck, the cost of travel on these links would not properly be represented as a function only of the flow on these links.

Unfortunately, phenomena of this type are very common in practice. Queues from a traffic intersection propagate back to another intersection; queues on a freeway propagate back to an upstream ramp. Although engineers concerned with traffic operations and control have analyzed specific problems under such headings as area signal control and freeway ramp control, there is a rather wide gap between the theorist who deals with traffic assignment and the practitioner who deals with the real world. A theory of traffic assignment, in which travel cost on a link depends on the flows on more than one link, has been developed under very restrictive assumptions that guarantee the existence of a generalization of the T^* for the user-optimal assignment (1).

(See, for example, reference 5.) However, this theory has so far been of limited use. On the one hand, to use the theory one must assign specific numerical values to the travel costs, and there is limited data with which to work; that is, the theory is more one of existence than practice. On the other hand, in some specific models involving queues on freeways, the solution of the user-optimal traffic assignment problem is not necessarily unique (see, for example, reference 6) and does not appear to be compatible with the type of generalization of the theory proposed in traffic assignment schemes.

6.6
Multiple Origins and/or Destinations, Costs Dependent on Flows

Instead of assuming that the c_{kl} are known (and independent of the flows), we now assume that c_{kl} is a known function $c_{kl}(f_{kl})$ of the flow f_{kl} on the link (k, l).

The assignments according to criteria (1) and (2) are, in principle, not much different from those discussed previously in sections 6.2, 6.3, and 6.4, except that there is no obvious way to evaluate the assignment explicitly; one must determine it by some scheme of successive approximations. For any reasonable size network, the calculations will generally consume considerable computer time.

The difficulty in generalizing the method for equalizing costs on various routes for a single origin and destination, as described in section 6.2, comes from the fact that these routes generally include links that carry flows between other origins and destinations. Any shift in flow to equate costs along routes between an O-D pair is likely to disrupt the balance of costs for routes between other O-D pairs. The difficulty in generalizing the methods of using cheapest paths described in section 6.4, is that one does not know the link costs (which are needed to compute the cheapest path) until one has found the link flows, but one cannot determine the link flows until one has determined the assignment.

Efficient computer algorithms will iteratively take one estimate of link costs, evaluate an assignment from these link costs, calculate the link flows, and then reevaluate the link costs. It will try to exploit the fact that trips between the same origin and destination will try to equate costs along various routes and the fact that the shortest-path algorithms are very efficient in selecting a few routes from an astronomical number of possible routes.

It is not obvious, with the link costs dependent on the flows, that there exists an assignment of type (1); that is, an assignment such that no trip can find a route that is cheaper than the one it has been assigned. It is conceivable that if one trip were to shift routes, it would set off a "chain reaction." It would increase the cost on the new route, which might cause trips using links of this route to choose other routes, which, in turn, would induce others to change. Finally, one might find that the original trip is displaced back to where it started. This does not happen for "sufficiently small" shifts in flow, but one can generate instability or oscillation if one tries to shift too much flow in one step.

To show there exists an assignment of type (1), consider the "objective function"

$$T^* = \sum_{(l, m) \in L} \int_0^{f_{lm}} c_{lm}(y) dy. \tag{6.22}$$

which is the obvious generalization of (6.15) to a network. We wish to minimize this with respect to the link flows f_{lm}, subject to the same constraints as applied to (6.21); that is, it must be possible to generate the f_{lm} from route flows satisfying (6.19).

The T^* is a continuous function of all the f_{lm} and is bounded from below ($T^* \geq 0$). If there exists any set of route flows satisfying (6.19) for which T^* is finite, then there must be an assignmont for which (6.22) assumes its minimum value with respect to all assignments that satisfy (6.19). We will now show that any assignment that minimizes (6.22) also satisfies the condition that no trip can find a cheaper route. Consequently, there are assignments of type (1).

For any admissible flow pattern, one can evaluate the f_{lm}, the $c_{lm}(f_{lm})$, and the cheapest route between any origin i and destination j. Suppose, for some assignment, that there is a route R_1 between i and j carrying a nonzero route flow and another route R_2 from i to j such that the cost on R_2 is less than on R_1. If we shift an infinitesimal flow ε from R_1

to R_2, T^* will change by

$$\Delta T^* = \sum_{(l,m)} \frac{\partial T^*}{\partial f_{lm}} \Delta f_{lm}$$

$$= \varepsilon \sum_{(l,m)\in R_2} c_{lm}(f_{lm}) - \varepsilon \sum_{(l,m)\in R_1} c_{lm}(f_{lm})$$

$$= \varepsilon[c(R_2) - c(R_1)] < 0.$$

If the original flow pattern was admissible, so is the new one, but the T^* for the new pattern is less than the original. Consequently, the original flow pattern does not minimize T^*. In order for any admissible flow pattern to minimize (6.22), it must be true that no trip can find a cheaper route from its origin to destination.

Because the $c_{lm}(f_{lm})$ are assumed to be monotone increasing functions of f_{lm}, it follows that all second derivatives of T^* with respect to f_{lm} are positive (for $f_{lm} > 0$). This, in turn, implies that T^* has no finite maxima or other stationary points except at its minimum. In the space of admissible link flows f_{lm}, there is a unique set of link flows that gives the minimum.

It is very useful in computations that there is a unique set of admissible link flows yielding the minimum T^*. However, this does not imply that there is a unique assignment of trips to routes or even a unique set of commodity flows. There are ways in which trips can trade parts of their routes, even trips from different origins and destinations, leaving the f_{lm} and, therefore, T^* unchanged.

If we wish to assign trips according to assignment (2), we should minimize

$$T = \sum_{(l,m)\in L} f_{lm} c_{lm}(f_{lm}) \tag{6.23}$$

subject to the same constraints as for (6.22). If we define the marginal link cost as

$$c_{lm}^*(f_{lm}) \equiv c_{lm}(f_{lm}) + f_{lm} dc_{lm}(f_{lm})/df_{lm}, \tag{6.24}$$

then we can write T in the form

$$T = \sum_{(l,m) \in L} \int_0^{f_{lm}} c_{lm}^*(y)dy. \tag{6.25}$$

This converts T into exactly the same form as T^*, except the c_{lm} have been replaced by c_{lm}^*. If we assume that the $c_{lm}^*(f_{lm})$ are monotone increasing functions, as with the $c_{lm}(f_{lm})$, then everything that was said about T^* has an obvious translation into a corresponding statement about T, with "cost" replaced by "marginal cost." In particular, a necessary condition for an optimal flow is that each trip choose a route of minimum total marginal cost between its origin and destination.

**6.7
Computation
Procedure**

Although there is not much difference in concept between the single O-D network and the multiple O-D network, computation procedures are quite different. The existence of an (implied) objective function for either type assignment guarantees that an iterative scheme of successive improvements can always be devised. The convexity of T or T^* implies that if one starts with any admissible "trial" solution and successively reassigns flows to better routes so as to guarantee that T or T^* decreases at each stage, one will never get trapped in a local minimum. Any procedure that corrects poor assignments will eventually converge to the optimal flow. If this were not the case, one could seriously question whether people in real life situations could assign themselves in a way approximating that of minimum cost for each trip. The only way travelers could hope to achieve an optimal route for themselves would be by a trial procedure. A computer scheme of successive improvements would thus mimic what travelers presumably would do. The only issue is to find a scheme of iteration that will converge rapidly and involve a manageable number of computations.

There is no escaping the fact that when flow on one route is shifted, it changes the travel times on all links of the route, which, in turn, can upset the balance of flows between other O-D pairs. Although one can be certain that a well-conceived scheme of improvement will converge, one cannot guarantee that any scheme will converge rapidly. In devising improvement procedures that shift sizeable flows at a time, one must also be careful not to "overshift." In the early stages of the improvement scheme, it is desirable to make shifts as large as possible (to shift only small increments at a time may require too many iterations), but in such a way as to make as large an improvement as possible. In later stages, one must be careful not to shift so much that one must shift back at a later stage and get caught in an induced oscillation. The best way to improve convergence, however, is to start with a good trial solution.

In practice one might use a procedure similar to the following:

1. Find the cheapest routes between all O-D pairs with known costs $c_{lm}(0)$ (which can be done quickly).

2. Take some suitable fraction of the total flow and assign it all to these shortest routes. The suitable fraction (perhaps 10 percent) should be chosen so that this assignment will not already create congestion on certain links and, hopefully, will not cause very large changes in the costs $c_{lm}(f_{lm})$. This is not an optimal assignment even for the fractional flow, but there is no reason for redistributing it at this stage.

3. Evaluate the $c_{lm}(f_{lm})$ using the flows assigned in step 2 and recalculate the shortest routes. One may find now that the optimal routes have changed because congestion on the old routes has made new routes cheaper. Now take a new fraction of the total flow and assign it to these optimal routes.

4. Repeat step 3 until all the flow has been assigned to routes.

The purpose of these steps is to create a reasonable trial solution. If one can keep track of where and why congestion is building up as the network is loaded then one can perhaps also devise special remedies. In the absence of any intermediate reassignments, however, one must load the network gradually to allow subsequent assignments to avoid congested links. There is no assurance that a more careful assignment of fractional flows will significantly improve the final trial solution. As one loads more flow, eventually one may find that the flows assigned in the early stages must be reassigned anyway.

Many of the early assignment algorithms stopped at step 4 with the expectation that if one loaded the network gradually, such an assignment would be accurate enough. Other algorithms recalculated new shortest paths and shifted some predetermined fraction of the previously assigned trips to the new shortest routes. There were possible convergence problems if one shifted too many trips at a time, but these reassignments were often iterated only a few times (to save computation time), again with the expectation that this would be accurate enough.

The more recent schemes also recalculate shortest paths, but, to prevent overshifting of flows, they treat the fraction of trips shifted to new routes as a variable. Using one parameter optimization schemes, they evaluate that fraction of the previously assigned trips and corresponding fraction of the link flows (either all trips or some subclassification of them) that, when shifted to the new routes, will minimize the T^* to within some predetermined accuracy. With these new flows on the links, the shortest paths are calculated again. The T^* is again minimized with respect to the fraction of flows shifted to these routes. The procedure is continued until the shifts are negligible according to some criteria. It is guaranteed to converge because T^* is decreasing at each iteration and cannot terminate as long as some trip can find a (sufficiently) better route.

This procedure, particularly the last step of successive improvement, may be quite long. It does, however, avoid the necessity of looking at all (or a large sampling of all) possible routes, which is clearly beyond the capability of any computer. Each iteration requires the reevaluation of costs on links and the evaluation of cheapest routes, which can be done rapidly. The only issue is the number of iterations. If the final step requires only 10, 100, or 1,000 steps, there is no problem; if it requires 10^6 iterations or more, one may be in trouble. As with most computer algorithms, one can invent problems for which convergence is so slow as to make the calculation impossible; but one hopes, with good judgment, to avoid such things. Unfortunately it is difficult to write a computer program that would identify and exploit all the

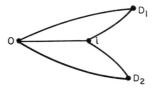

Figure 6.7
Traffic assignment between a
single origin and two destinations.

things that a person doing hand calculations would recognize immediately.

6.8
An Example

This theory is abstractly very elegant, but the computations are rather tedious, especially considering that the conclusions cannot be taken too seriously. Perhaps if one understood better some of the qualitative aspects of typical or troublesome situations, one could avoid using a computer to calculate the obvious and find special algorithms to handle the troublesome situations.

The simple network shown in figure 6.7 illustrates why, in more complex networks, it may require many iterations to arrive at a solution. Suppose one has a single origin and two (or more) destinations, and route OiD_1 has a link Oi in common with a route OiD_2. There are also alternative routes OD_1 and OD_2 not through i. Suppose also that the cost functions on all links are known and that one has achieved an optimal assignment with a flow q_{OD_1} between O and D_1 and a flow q_{OD_2} between O and D_2. If this assignment gives a nonzero flow to both OiD_2 and OD_2, then the cost from O to D_2 is independent of the path; similarly for the flow from O to D_1. Now suppose one adds some additional flow from O to D_2. It will distribute itself so as to maintain the equality of cost by way of OiD_2 and OD_2. Some of the flow will certainly be assigned to OiD_2, thereby increasing the cost on Oi. This, in turn, will force some of the flow on OiD_1 to take OD_1 and increase the travel cost from O to D_1 by way of either route.

It is impossible to avoid this complication by loading the two O-D pairs simultaneously. The route OD_2 may be quite congested already at some flow level q_{OD_2}. Anything one adds to q_{OD_2} is assigned mostly to OiD_2. Even though one adds flow simultaneously to q_{OD_1}, one may find that the flow on route OiD_1 must decrease. Whereas a flow could be assigned to route OiD_1 in the early loading, the final loading may drive it completely away.

This example also shows that as one adds more trips to a network, the flow on some links may decrease. In particular, the flow between i and D_1 decreases. The travel time from O to D_1 and from O to D_2 both increase, however, regardless of where new flow is added between O and D_1 or O and D_2. Generally an increase in the flow of any single commodity will increase the costs for that commodity but will not necessarily increase the cost for other commodities. In particular, if i is an origin and D_1 the destination for some trips, an increase in the trips from O to D_2 decreases the flow on iD_1 and, therefore, decreases the cost of travel for trips that must be made from i to D_1.

Problems   1

On a network of one-way streets $(0, 1)$, $(2, 0)$, $(2, 1)$, a flow q_{01} wishes to go from 0 to 1 and a flow q_{21} wishes to go from 2 to 1. If the travel cost on each of the links, $c_{01}(f)$, $c_{20}(f)$, $c_{21}(f)$, and their corresponding marginal costs are increasing functions of f, how would one evaluate the flows on the various links if
a everyone chose the cheapest route to himself;
b. traffic is assigned so as to minimize the total travel cost; or
c. traffic is assigned so as to minimize the cost to the flow q_{21} only?

2

There are two routes from an origin O to a destination D with no common links. Two types of users, A and B, with flows f_A and f_B, wish to go from O to D. The cost of travel on route j ($j = 1, 2$) is $c_j(f_j)$ for a flow f_j on route j; $f_j = f_{Aj} + f_{Bj}$, and f_{Aj} and f_{Bj} are the flows of type A and B on route j.

Type A users (taxis, for example) wish to minimize the total cost of travel for all of f_A. Type B users (private cars) wish to minimize their individual costs. Neither A nor B is concerned with costs to the other.

How will the flows $(f_{A1}, f_{A2}, f_{B1}, f_{B2})$ be distributed between the routes?

3

A route (or collection of many routes) between an origin O and destination D, with (collective) cost function $c_1(f)$, carries a flow f from O to D. A consultant is asked to recommend whether it is worthwhile to build another route of specified type from O to D. The new route would cost A to build and would have a trip cost $c_2(f_2)$. It is assumed that travelers would choose the route with minimum cost to them. The consultant compares total travel cost savings with construction cost, concludes that the saving in travel cost exceeds the construction cost, and, therefore, recommends that the new route be built. (The costs are all costs per year under static conditions; the construction cost really is interest, maintenance, and so forth.) Despite the recommendation, the facility is not built. A few years later, another study is authorized. Meanwhile the flow has increased from f to f^*, $f^* > f$, but there has been no inflation; all costs are the same as before. Is it possible that the consultant could now recommend, on the basis of the same arguments as before, that the facility *not* be built? Assume that all costs and marginal costs are increasing functions of f.

References   The "modern" interest in the traffic assignment problem can be considered to have started with the following paper:

1
Wardrop, J. G. "Some Theoretical Aspects of Road Traffic Research," *Proceedings of the Institute of Civil Engineering.* Part II, 1 (1952): 325–378. Wardrop discusses the two types of assignment principles (minimize individual or total costs) as applied to a single origin–single destination. The two principles have since been identified as "Wardrop's first and second principles" (although there is frequent confusion as to which is the "first" and which the "second.")

The traffic assignment problem as applied to just two routes dates back (at least) to 1920:

2
Pigou, A. C. *The Economics of Welfare*, 1st ed. New York: Macmillan, 1920; 4th ed., 1932.

For a rather thorough discussion of the assignment problem and some of its economic implications, see

3
Beckmann, M., McGuire, C. B., and Winsten, C. B. *Studies in the Economics of Transportation.* Yale University Press, 1956.

Some discussions of various travel cost functions are contained in

4
Branston, D. "Link Capacity Functions: A Review," *Transportation Research* 10 (1976): 223–236.

5
Dafermos, S. C. "An Extended Traffic Assignment Model with Application to Two-Way Traffic," *Transportation Science* 5 (1971): 366–389.

6
Newell, G. F. "The Effect of Queues on the Traffic Assignment to Freeways," *Proceedings of the Seventh International Symposium on Transportation and Traffic Flow Theory* (Kyoto, Japan, 1977), pp. 311–340.

For a systematic mathematical analysis of the general assignment problem, see

7

Dafermos S. C., and Sparrow, F. T. "The Traffic Assignment Problem for a General Network," *Journal of Research. National Bureau of Standards* 73B (1969): 91–118.

8

Nguyen, S. "An Algorithm for the Traffic Assignment Problem," *Transportation Science* 8 (1974): 203–216.

9

Steenbrink, P. A. *Optimization of Transport Networks*. New York: John Wiley & Sons, Inc., 1974.

10

Florian, M. A., ed. *Traffic equilibrium methods*. Lecture Notes in Economics and Mathematical Systems. No. 118. Berlin: Springer-Verlag, 1976.

7 ASSIGNMENTS ON IDEALIZED NETWORKS

7.1
Introduction

We consider again some of the analytic methods introduced in chapter 4 in order to show how the analysis of the traffic assignment to various simple networks can illustrate important issues in transportation. Because the advantage of analytic techniques lies primarily in their ability to demonstrate how some objective (for example, the total cost of transportation) depends on various parameters (which, in a numerical scheme, could only be studied one at a time), the examples chosen here will be aimed particularly at showing how the flows on a network depend on certain "input variables" to the traffic assignment, namely the travel cost functions, the network geometry, and the O-D table.

The range of problems that can be analyzed by analytic techniques is obviously very limited. One must almost manufacture examples to illustrate certain points. It may therefore be difficult to see how some of these unrelated examples can be incorporated into any general scheme for analyzing real networks. On the other hand, one cannot make intelligent decisions as to what are reasonable proposals for a present or future transportation system without understanding the qualitative consequences of various actions.

The types of simple networks that can be analyzed by analytic methods are mainly those involving only a few nodes and links or those containing a very large number of nodes and links with a very simple geometry (such as those introduced in chapter 4).

7.2
Improvements in a
Network, Single
O-D

When an improvement is made in a network (such as a road is built or widened), the improvement often represents only a small change in the existing network. One often hears the complaint from drivers that a new freeway is congested the day it is opened. In many situations, this is what should happen.

To illustrate this, suppose that intially there are n identical routes between a single origin O and a single destination D, each with a cost function $c_1(\cdot)$. A total flow q between O and D is equally divided among the routes; the flow f_k on the kth route is q/n; and the common cost on all routes is $c = c_1(q/n)$.

Suppose that an improvement consists of building one more route, changing n to $n + 1$, but the total flow q from O to D remains the same. The new flows are $q/(n + 1)$, and the cost per trip is $c_1(q/(n + 1))$. If $n \gg 1$, the addition of one more route does not change the flow very much on the old routes, nor does it change c very much. The new route takes a little flow off each of the old routes. The flow on the kth route changes by

$$\Delta f_k \simeq -q/n^2 \simeq -f_1/n,$$

and the cost per trip decreases by

$$\Delta c \simeq \frac{dc_1(f_1)}{df_1} \Delta f_k \simeq -\frac{dc_1(f_1)}{df_1}\frac{f_1}{n}.$$

All travelers receive the same small benefit; the travelers who move to the new route gain no more than those who remain on the old routes. Because most of the trips remain on the old routes, most of the benefit comes from the reduction in cost for the trips that do not change routes (contrary to the naive assumption that the benefit of a new facility derives from those who use it). Although each traveler thinks that the improvement is very small (and complains that "things are just as bad as they were before"), the total change in cost for all trips is

$$\Delta T = q\Delta c \simeq -q\frac{dc_1(f_1)}{df_1}\frac{f_1}{n} = -f_1^2\frac{dc_1(f_1)}{df_1}, \qquad (7.1)$$

which does not depend on n.

The geometrical interpretation of ΔT can be seen in figure

6.4. If one draws a tangent line to the curve $c_1(f_1)$ and extends it back to $f_1 = 0$, a rectangle of height $f_1 dc_1(f_1)/df_1$ and base f_1 is formed. The ΔT in (7.1) is the area of this rectangle. If there is little congestion on the original routes, then $dc_1(f_1)/df_1$ is small. Obviously there is little benefit to be gained from building a new route if it will not change the cost per trip.

One can easily generalize this argument to nonidentical routes. Suppose that initially we had n routes with cost functions $c_k(\cdot)$ for $k = 1, 2, \ldots, n$ and we added a new route with cost function $c_{n+1}(\cdot)$. Suppose also that all the original routes were used ($f_k > 0$) and that the new route now attracts only a small fraction of the total flow q. Initially the flows f_k are such that

$$c_k(f_k) = c$$

for some number c. The result of adding a new route is that each of the flows f_k changes by an amount Δf_k so as to give a slightly different cost $c - \Delta c$, where

$$-\Delta c = (\Delta f_k)dc_k(f_k)/df_k, \qquad k = 1, 2, \ldots, n.$$

But the flow taken from the routes must go on the new route,

$$f_{n+1} = -\sum_{k=1}^{n} \Delta f_k = \Delta c \sum_{k=1}^{n} [dc_k(f_k)/df_k]^{-1},$$

and the cost on the new route must also become $c - \Delta c \simeq c$,

$$c_{n+1}(f_{n+1}) = c, \qquad f_{n+1} = c_{n+1}^{-1}(c). \qquad (7.2)$$

The total benefit now becomes

$$\Delta T = q\Delta c = f_{n+1} \frac{(1/n) \sum_{k=1}^{n} f_k}{(1/n) \sum_{k=1}^{n} [dc_k(f_k)/df_k]^{-1}}. \qquad (7.3)$$

Both numerator and denominator of (7.3) have been multiplied by $1/n$ because the numerator can now be interpreted as the average flow per route, and the denominator can be

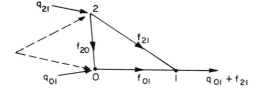

Figure 7.1
Freeway ramp control.

interpreted as the average of the reciprocal slopes of the $c_k(\cdot)$. The value of f_{n+1} in (7.3) is evaluated from (7.2). Of course, (7.3) reduces to (7.1) if the routes are identical. From (7.3) one sees that if *any* route is uncongested, $dc_k(f_k)/df_k \approx 0$ for any k, then $\Delta T \approx 0$. Otherwise, (7.1) shows more clearly than (7.1) that the benefit is proportional to the flow f_{n+1} removed from the old routes, with a coefficient depending on the congestion on the old routes.

7.3
Traffic Control,
Braess's Paradox

Consider the simple three node network (introduced in problem 1, chapter 6), shown by the solid lines of figure 7.1, involving one-way links (0, 1), (2, 0), and (2, 1) in which a flow q_{01} seeks to travel from 0 to 1 and a flow q_{21} seeks to travel from 2 to 1. The flow q_{01} has only one possible route, but q_{21} has a choice between the routes (2, 0) (0, 1), or (2, 1). Suppose travel costs on the links, $c_{01}(\cdot)$, $c_{20}(\cdot)$, and $c_{21}(\cdot)$, are increasing and convex, and each traveler from 2 to 1 chooses the cheaper route.

We wish to know what will happen if (at negligible cost) the link (2, 0) is made worse in the sense that $c_{20}(f)$ is replaced by a new cost $c'_{20}(f)$ for which

$$c'_{20}(f) \geq c_{20}(f), \qquad f > 0.$$

For example, one could make $c'_{20}(f)$ infinite by closing the link. In particular, we want to know under what conditions it is possible for the total travel cost of all trips, T, to *decrease* when c_{20} is replaced by c'_{20}.

The fact that it may be possible, with a user-optimal assignment, for the total cost of travel in a network to increase when a new link is added to the network has come to be known among network analysts as "Braess's paradox" (see references 1 and 2). It was first discussed for a slightly more complicated network than in figure 7.1 consisting of two disjoint routes between the same origin and destination joined by a new link to provide a crossover connection between the routes. Figure 7.1 can be modified so that the flows q_{01} and q_{21} reach points 0 and 2 along two links from a common origin as indicated by the broken lines.

If it is possible to decrease the total travel cost by making a link worse or removing it, it obviously must also be possible to increase the travel cost by improving a link or by adding a link. That this should be called a "paradox" adds intrigue to something that, in principle, is well known to traffic engineers. Any time one restricts the flow on a street for the purpose of "traffic control," one is, in effect, increasing the cost of travel on that link, presumably in order to improve the flow on the network as a whole.

For the most popular current interpretation of figure 7.1, imagine that the link (0, 1) is a section of a freeway and q_{01} represents the through traffic. The link (2, 0) is one entrance ramp, whereas the link (2, 1) is a schematic representation of an alternate route onto the freeway by way of a second downstream ramp. The question now becomes: Under what conditions is it advantageous to regulate the flow or close a ramp? Regulating the flow ("ramp metering") is equivalent to inserting a bottleneck on this link with some specified value of the f_0 in figure 6.6. If, with no regulation, a flow f_{20} exceeding f_0 were to use this link, then a queue would form on the ramp. As explained in section 6.5, the queue will grow until the travel time on the routes (2, 0), (0, 1), and (2, 1) are equal with a flow f_0 on (2, 0), provided that the ramp has sufficient storage for this to happen.

In chapter 6 we discussed the consequences of imposing tolls as a means of traffic control. To the individual traveler, the imposition of a toll on the link (2, 0) is equivalent to making this link worse; that is, $c_{20}(f)$ changes to a $c'_{20}(f)$. In chapter 6, however, the revenue collected from the tolls was considered a benefit. The net transportation cost was still defined by the original cost $c_{20}(f)$. The cost difference $c'_{20}(f) - c_{20}(f)$ was considered as simply a transfer of funds from the individual traveler to society (at no loss). Now we, in effect, assume that the collection of appropriate tolls is not feasible. If we make a link worse (for example, by allowing a queue to form), then the extra expense incurred by anyone who continues to use that link cannot be recovered. Delay due to queueing is a pure loss.

Finally, to answer the question posed, we note first that if the marginal cost on route (2, 0), (0, 1), even for $f_{20} = 0$, is larger than the marginal cost on route (2, 1) at the flow q_{21},

$$c_{20}(0) + c_{01}^*(q_{01}) > c_{21}^*(q_{21}), \tag{7.4}$$

then the assignment of all q_{21} to the route (2, 1) minimizes total cost. This is not necessarily the assignment that minimizes the cost to each traveler. It may be that

$$c_{20}(0) + c_{01}(q_{01}) < c_{21}(q_{21}). \tag{7.5}$$

This might be true if, for example, (0, 1) is a congested freeway with a high marginal cost, $c_{01}^*(q_{01})$, but a competitive individual cost $c_{01}(q_{01})$. The optimal assignment obtains if one closes the link (2, 0) completely. (This is done on some freeways under such conditions, particularly if the freeway link (0, 1) has a bottleneck and q_{01} is close to the capacity of the bottleneck).

Because closing the ramp is equivalent to making $c_{20}(f_{20}) = \infty$ (with an ill-defined cost, $f_{20}c_{20}(f_{20}) = 0 \cdot \infty$), one might ask what would happen if a toll were charged so as to make f_{20} arbitrarily small or if the ramp were metered at an arbitrarily low rate. In either case, the cost of travel $c_{20}(f_{20})$ on the link (2, 0) would remain finite as the toll or metering rate caused f_{20} to approach zero. The total cost of travel on the link (2, 0), $f_{20}c_{20}(f_{20})$ would, therefore, go to zero as $f_{20} \to 0$. It makes no difference how f_{20} goes to zero. In particular, if one charged tolls, one would collect zero revenue.

Suppose now that (7.5), but not (7.4), is true, so that it is advantageous to both society and individual travelers that $f_{20} > 0$. However, suppose the user-optimal assignment, with equal costs on the two routes (or with $f_{20} = q_{21}$), produces a flow f_{20} in excess of the social optimal assignment. If, to discourage travel on $(2, 0)$, one simply increases the cost of travel on $(2, 0)$, the flow f_{20} that remains will pay a higher cost, and this cost should be added to the total cost.

The total cost of travel is now

$$
\begin{aligned}
T' &= f_{20}c'_{20}(f_{20}) + (q_{01} + f_{20})c_{01}(q_{01} + f_{20}) \\
&\quad + (q_{21} - f_{20})c_{21}(q_{21} - f_{20}) \\
&= q_{01}c_{01}(q_{01} + f_{20}) + q_{21}c_{21}(q_{21} - f_{20}) + f_{20}[c'_{20}(f_{20}) \\
&\quad + c_{01}(q_{01} + f_{20}) - c_{21}(q_{21} - f_{20})].
\end{aligned} \tag{7.6}
$$

Under a user-optimal assignment with $f_{20} > 0$ and $f_{21} = q_{21} - f_{20} > 0$, travelers would force the costs to be equal on the two routes; consequently, the last term of (7.6) vanishes and

$$
T' = q_{01}c_{01}(q_{01} + f_{20}) + q_{21}c_{21}(q_{21} - f_{20}). \tag{7.7}
$$

This equation simply says that all trips q_{01} pay a cost $c_{01}(q_{01} + f_{20})$ and all trips q_{21} pay the cost $c_{21}(q_{21} - f_{20})$ on link $(2, 1)$ whether they use it or not. Notice that (7.7) does not contain the cost function $c'_{20}(\cdot)$ explicitly but contains f_{20}, which can be changed through a modification of $c'_{01}(\cdot)$.

Even for small values of f_{20}, there is no guarantee that

$$
\begin{aligned}
\frac{dT'}{df_{20}} &= q_{01}\frac{dc_{01}(q_{01} + f_{20})}{d(q_{01} + f_{20})} - q_{21}\frac{dc_{21}(q_{21} - f_{20})}{d(q_{21} - f_{20})} \\
&\rightarrow q_{01}\frac{dc_{01}(q_{01})}{dq_{01}} - q_{21}\frac{dc_{21}(q_{21})}{dq_{21}}, \quad f_{20} \rightarrow 0
\end{aligned} \tag{7.8}
$$

is negative. Thus it is not necessarily advantageous to allow any flow to use the link $(2, 0)$ if the only means of control is to make $c'_{20}(\cdot)$ so large that the costs become equal on the two routes.

Equation (7.8) also has a simple interpretation. The first term is the cost imposed on others by adding one traveler to the link $(0, 1)$, and the second term is the savings to other travelers if one traveler is removed from link $(2, 1)$. The cost to the traveler who shifts is irrelevant because the costs of travel on the two routes have been made equal.

The costs T and T' are equal for $f_{20} = 0$ (as explained earlier), and they are again equal at a flow f_{20}^* such that

$$c_{20}(f_{20}^*) + c_{01}(q_{01} + f_{20}^*) = c_{21}(q_{21} - f_{20}^*);$$

that is, they are equal at the uncontrolled user-optimal assignment (provided $q_{21} > f_{20}^*$) because no control is necessary for this flow, and $c'_{20}(f_{20}) = c_{20}(f_{20})$. For $f_{20} > f_{20}^*$, $T' < T$, but it would be necessary for $c'_{20}(f_{20}) < c_{20}(f_{20})$.

For any f_{20} with $0 < f_{20} < f_{20}^*$, $T' > T$. If, however, $q_{21} < f_{20}^*$, so that all trips use $(2, 0)$ under a user-optimal assignment because route $(2, 0)$, $(0, 1)$ is definitely cheaper at $f_{20} = q_{21}$, there are some other interesting cases. One certainly would not regulate the flow f_{20} so as to decrease f_{20} only slightly below q_{21} because, to initiate any control, the cost on link $(2, 0)$ must first be raised to a value so that the costs are equal on the two routes; that is, T must be increased to T' at $f_{20} = q_{21}$. Any benefits from a decrease in f_{20} must overcome this initial cost. (See reference 3 for a more thorough discussion of these cases.)

7.4 Symmetry

Many idealized network flow problems can be solved by exploiting their symmetry. Suppose for a given network there exists a permutation of the nodes of a graph that maps the graph into itself. For example,

an equilateral triangle or square maps into itself by a cyclic permutation of the nodes;

a ring radial network maps into itself by a rotation about the center;

a square grid maps into itself under a translation by a lattice spacing along either grid direction; and

a square grid of two-way links maps into itself under reflection about any grid line or under a rotation of the plane through 90°.

If there is a traffic assignment problem associated with such a network, suppose also that under this mapping of the network into itself the complete traffic assignment problem maps into the same problem; that is, the image of any link under the mapping has the same travel cost function as the original link. Futhermore, if some node O maps into O' and node D maps into D', then the flow from O' to D' is the same as from O to D.

In the mathematics literature, the collection of all mappings that transform a space (graph, network) into itself is known as a *group*. If Γ_1 and Γ_2 are two such transformations, then the mapping Γ_1 followed by Γ_2 also transforms the space into itself and is a member of the group. This mapping is denoted by $\Gamma_2\Gamma_1$. The formal definition of a group Ω is a collection of objects $\Gamma_1, \Gamma_2, \ldots$ with the following properties:

If Γ_1 and Γ_2 are elements of Ω, there is an element of Ω associated with $\Gamma_2\Gamma_1$.

There is an identify $I \in \Omega$ for which $I\Gamma = \Gamma I = \Gamma$ for all $\Gamma \in G$.

Every Γ is associated with an element Γ^{-1}, the inverse of Γ, for which $\Gamma\Gamma^{-1} = \Gamma^{-1}\Gamma = I$.

The importance of symmetry in the present context arises from the fact that if traffic is assigned to routes so as to minimize some convex function of the link flows (such as the T or T^* described in chapter 6), then the resulting link flows must show the same symmetry as the network itself. The actual route assignments are not necessarily unique, and it is not necessarily true that every optimal route assignment shows the symmetry of the network, but one can show that some possible optimal route assignment does have this symmetry. Thus a symmetric problem has a symmetric solution.

To prove these properties, suppose a traffic assignment problem with a certain symmetry has an optimal link or route flow pattern that is not symmetric. For example, in figure 7.2a, a square network of nodes 1, 2, 3, 4 and undirected links (1, 2), (2, 3), (3, 4), (4, 1) all have the same travel cost functions $c(f)$. The O-D table has q trips from 1 to 3, 3 to 1, 2 to 4, and 4 to 2. One possible assignment is to route the trips as in figure 7.2b, with trips between 1 and 3

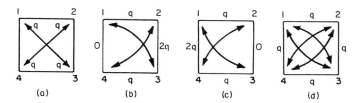

Figure 7.2
Routing of traffic on a square
network.

going by way of the route 1-2-3, and those between 2 and 4
going by way of the route 2-3-4. This gives a two-way flow
of q on links (1, 2) and (3, 4), $2q$ on (2, 3), and 0 on (4, 1).

If the flow pattern is not symmetric, there exists some
transformation Γ that maps the network into itself but maps
the present flow pattern into a different flow pattern. Thus,
in figure 7.2, a reflection or a rotation through 180° maps
the flow pattern of figure 7.2b into 7.2c. It is assumed,
however, that the symmetry includes symmetry not only of
the network geometry but also the cost functions, O-D table,
and the objective function. If Γ produces a new flow pattern,
this pattern must be an optimal solution of the transformed
assignment problem, which, in turn, is identical to the
original problem. We thus have two different optimal flow
patterns; that is, they have equal values for the objective
function.

According to the results of chapter 6, however, the
objective functions T or T^* (or any convex function of the
link flows) have a unique set of optimal link flows. Thus the
transformed link flows cannot yield a different optimal link
flow pattern than the original; the link flows must be
invariant to the symmetry transformations. In figure 7.2, the
only possible symmetric flow pattern is shown in figure 7.2d
for which the trips from 1 to 3 split, half going by way of
1-2-3 and half going by way of 1-4-3; similarly for the other
O-D pairs.

The convexity of the objective function is important, for
suppose that the costs $c(f)$ were decreasing functions of f;
that is, suppose there were an "economy of scale." The total
cost of the assignment 7.2b or 7.2c would be less than 7.2d if

$$(2q)c(2q) < 2[qc(q)];$$

that is,

$$c(2q) < c(q).$$

Indeed, under these conditions (b) and (c) would be the least-cost assignments; they would not be unique, and a symmetric problem would not have a symmetric solution (except in the sense that all transformations of an optimal assignment are also optimal). Assignment (d) would be uniquely most expensive.

Even if the objective function is convex so that the optimal assignment has a unique symmetric link flow pattern, the route flow pattern need not be symmetric. It is clear, for example, that if we did have a symmetric route flow pattern in which several routes crossed, it would be possible for trips to exchange sections of their routes in an asymmetric way so as to create an asymmetric route flow pattern with the same optimal link flows.

To show that there are some symmetric optimal route flows, suppose we start with an optimal route flow pattern that is not symmetric (but gives the optimal symmetric link flows). Consider, for example, the network shown in figure 7.3a having q trips from 1 to 3, 3 to 1, 3 to 6, 6 to 3, 1 to 6, and 6 to 1. One possible route assignment would be to send the trips between 1 and 6 by way of the route 1-2-3-7-6, those between 1 and 3 by way of 1-4-3, and those between 3 and 6 by way of 3-5-6, as shown in figure 7.3b. This route assignment is not symmetric to reflections across the diagonal 1-3-6. If we reflect the assignment (b), we obtain the one shown by (c).

For any transformation Γ of the symmetry group and any assignment, such as (b), we can construct a new route flow assignment by taking half of every route flow in the original assignment (b) and adding half of every route flow in the mapping of this assignment under Γ, as shown in assignment (c). This linear combination of route flows will satisfy the O-D table and the conservation equations. If the original route flows gave the unique optimal link flows, so do the

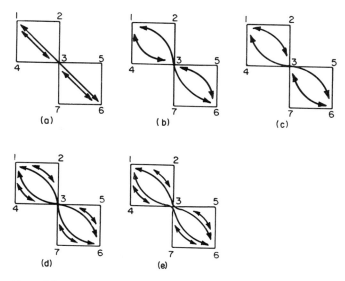

Figure 7.3
Asymmetric route flows with
symmetric link flows.

transformed route flows and the mixture of the two. This
new route flow pattern, however, will be invariant to
transformations by Γ. The route flow pattern generated from
(b) and (c) is shown in (d). The symmetric route flow pattern
may not be unique. The route flow pattern shown in (e),
having route flows along 1-2-3-5-6 and 1-4-3-7-6 is
symmetric, as is any linear combination of those shown in
(d) and (e).

More generally, if the symmetry group can generate a
total of n different route assignments from a given optimal
route assignment, a new route assignment can be created by
superimposing $1/n$ times each of these assignments. This
flow pattern will be symmetric under all transformations of
the group. If one wishes to determine an optimal route flow
assignment on a network having some symmetry and a
convex objective function, it is sufficient to consider only
symmetric route assignments.

**7.5
Rectangular Grid**
As a more interesting illustration of symmetry principles,
suppose we have a rectangular grid of two-way roads. The

cost of travel on any link between lattice points may depend on the flow on that link (or even on the flows on neighboring links), but the cost function does not depend on the location of the link in the grid. Thus there are just four (possibly equal) cost functions, $c_N(\cdot)$, $c_S(\cdot)$, $c_E(\cdot)$, and $c_W(\cdot)$, for travel of one block length in the north, south, east, and west directions, respectively.

Suppose that trips have origins and destinations at the lattice points. If the roads have spacing a in the E-W direction, b in the N-S direction, then the lattice points have rectangular coordinates $[ia, jb]$, $-\infty < i, j < +\infty$, relative to some arbitrarily chosen origin. If we measure E-W distances in units of length a and N-S distances in units of b, the lattice points may also be labeled simply as $[i, j]$.

The O-D table specifies the number $q([i, j], [i', j'])$ of trips from an origin with grid coordinates $[i, j]$ to a destination $[i', j']$. This is assumed to have translational symmetry in both the N-S and E-W directions; that is,

$$q([i, j], [i', j']) = q([0, 0], [i' - i, j' - j]) \equiv q(i' - i, j' - j).$$
(7.9)

The complete O-D table is uniquely determined by the number of trips from any origin with trip length components $i' - i, j' - j$.

The rectangular grid is assumed to extend infinitely far, but if one is worried about boundary conditions at ∞, one can create an artificial network in which the rectangular grid is wrapped on a torus (doughnut) so that the network has cyclic symmetry in two directions.

The network, the O-D table, link travel costs, and any objective function are assumed to be invariant to translations by a unit distance in either grid direction. We conclude from this that the link flows have the same symmetry; that is, there are, at most, four different (as yet unknown) flow values f_N, f_S, f_E, f_W in the four direction. Each link has the same flow as any parallel link.

Any trip from $[i, j]$ to $[i', j']$ must travel at least a distance $i - i'$ along E-W links and $j - j'$ along N-S links. For an optimal flow pattern, it is obvious that no trip will take a longer path than this minimum, nothing can be gained by generating unnecessary travel. We do not know the actual routes, but they are not unique.

To evaluate the flows f_N, f_S, f_E, f_W, one may assign the flow from $[i, j]$ to $[i', j']$ to any minimum distance route, provided that for any $[k, l]$ one assigns the flow from $[i + k, j + l]$ to $[i' + k, j' + l]$ to the corresponding translation of this route. Suppose, for example, that all trips first move in the E-W or W-E direction and then in the N-S or S-N direction. The flow f_E on the link from $[0, 0]$ to $[1, 0]$ will include all trips from $[i, 0]$ to $[i', j']$ ($i \leq 0$ and $i' \geq 1$, all j').

It is convenient, first, to recognize that in the calculation of f_E one need not know the distribution of the N-S component j'. The f_E depends only on

$$q_E(i) = \sum_{j=-\infty}^{+\infty} q(i, j), \qquad i > 0, \tag{7.10}$$

which represents the total trips from any origin to all destinations having a W-E component of the trip of i units, $i > 0$.

Any trip starting from $[k, 0]$, $-i < k \leq 0$, and traveling eastward a distance i will traverse the link $[0, 0]$ to $[1, 0]$. There are $iq_E(i)$ such trips for each i; therefore

$$f_E = \sum_{i=0}^{\infty} iq_E(i). \tag{7.11}$$

Similarly, we can define

$$q_W(i) = \sum_{j=-\infty}^{+\infty} q(-i, j), \qquad i > 0,$$

for trips in the westward direction, and represent f_W as

$$f_W = \sum_{i=0}^{\infty} iq_W(i).$$

For the N-S directions, the corresponding quantities are

$$q_N(j) = \sum_{i=-\infty}^{+\infty} q(i, j), \qquad q_S(j) = \sum_{i=-\infty}^{+\infty} q(i, -j), \qquad j > 0;$$

$$f_N = \sum_{j=0}^{\infty} jq_N(j), \qquad f_S = \sum_{j=0}^{\infty} jq_S(j).$$

Note that these calculations of link flows are essentially the same as those described in section 5.5 for a single route. In particular, see equation (5.23).

We have obtained the flows by a "direct" method of postulating an assignment and then evaluating the number of trips passing any point. A simpler indirect method is to recognize that if there were a flow f_E in the eastward direction and block length a miles, then there would be af_E vehicle miles of travel per block in the eastward direction. If the region in question contained N nodes and, therefore, also N eastward links, the total vehicle miles of travel in the eastward direction would be $f_E aN$.

The total vehicle miles of travel in the eastward direction, however, can also be written as the sum of the vehicle miles of travel made by all trips in that direction. A trip from origin $[0, 0]$ to destination $[i, j]$ travels east a distance ia. The O-D flow $q(i, j)$ from $[0, 0]$ to $[i, j]$ travels $iaq(i, j)$ vehicle miles eastward from $i > 0$. The sum of this over all j is $iaq_E(i)$. All trips from $[0, 0]$ contribute

$$a \sum_{i=0}^{\infty} iq_E(i)$$

vehicle miles of eastbound traffic. Because there are N origins, the total vehicle miles of travel in the eastward direction in the region is also given by

$$Na \sum_{i=0}^{\infty} iq_E(i) = f_E aN, \qquad (7.12)$$

which agrees with the result in (7.11). The flows f_W, f_N, f_S can be interpreted similarly.

Having found the f_E, f_W, f_N, f_S directly from the O-D table, the link travel costs can also be evaluated, $c_E(f_E)$, $c_W(f_W)$, $c_N(f_N)$, $c_S(f_S)$.

The example of the rectangular grid is very special. There are not many examples of ideal networks with symmetry in two directions; the only other examples would be networks with triangular, hexagonal, etc., grids. The existence of translational symmetry in two directions in a two-dimensional problem reduces the assignment problem to a trivial exercise, in effect, a zero-dimensional problem: Evaluate four numbers $f_E, \ldots, f_S$. (If we had added some reflection or rotational symmetry, this could have been reduced to two or one number.) Considered in the context of chapter 6, however, this solution shows a number of interesting features.

In general networks, the traffic assignment depends on the travel costs on links, which depends on the link flows, which, in turn, depends on the assignment. The evaluation of the link flows usually involves rather lengthy iterative calculations. If one tried to evaluate the flows on a square grid network (or a finite version of it) using conventional algorithms, and one had used any asymmetric scheme of loading the network so that at some stage of the calculation the flows were not symmetric, the algorithms would likely run for a very long time, shifting flows back and forth between nearly equal routes trying to make the link flows more nearly equal on parallel routes (but never quite succeeding).

The simple observation that the flows would be equal on all parallel links (or, more to the point, that the travel costs would be equal on parallel links), even if initially unknown, would break up this complicated dependence of everything on everything else. An optimal route assignment could be proposed even before the link costs were evaluated. The argument leading to (7.12) further demonstrates that the link costs are insensitive to the detailed route assignments. To evaluate the flow on any eastbound link, for example, it is necessary only to know the vehicle miles of travel in the eastward direction per square mile and distribute it equally on all links. It is also important to notice that the link flows are insensitive to trip length distributions; they depend only on the first moments of the distribution. Much of the detailed information contained in the O-D data (which is obtained at great expense) is, therefore, irrelevant.

The most important features of this solution are the things that it does *not* contain, the many parameters that are not relevant (and which are, perhaps, also not particularly relevant for more general networks). For this symmetric network, the flows do not even depend on whether traffic was assigned to minimize individual or total cost. The equality of individual costs and marginal costs on parallel routes are realized simultaneously. (The issue of assignment according to the two schemes is significant only when two facilities share trips with different cost functions.) Each link cost could even depend on several link flows (as would be desirable at a traffic intersection), provided that the optimal link flow assignments are unique and, consequently, have the symmetry of the network.

From a practical point of view, real networks do not have perfect symmetry and are not infinitely large; so the significance of this evaluation is dependent on the extent to which the idealized network approximates a real one. There are real networks for which certain finite regions have nearly rectangular grids. If the regions were sufficiently large, the idealization should be valid except near the boundary. "Sufficiently large," however, is a relative thing. There are two "natural" units of length in the problem, the grid size and the average trip length. Certainly a region of dimension large compared with a typical trip length would be sufficiently large to use these formulas.

For a region of dimension large compared with a block length but small compared with some trip lengths, one cannot deduce the flows f_l in this manner using the total vehicle miles of travel from the O-D table, but one still should expect the flows to be nearly equal on all parallel streets. The problem is to estimate the f_l, which are the total vehicle miles of travel in the region. One possible scheme would be to make a preliminary traffic assignment over the whole network, as in chapter 6. This may assign excessive and unrealistic traffic to some streets but realistic amounts to regions. Now the total vehicle-miles of travel can be calculated in the region and redistributed equally over all streets. If necessary, the costs $c_l(f_l)$ can be evaluated and used to reevaluate the assignment over the whole network.

Perhaps most important is the fact that many of the properties of this solution are quite insensitive to the symmetry of either the trip distribution or the network.

7.7 Modification of a Rectangular Grid

In section 7.2 it was argued that, for a network containing many routes, a new route or improvement of an existing route is not likely to change significantly the travel cost for an individual, but there is a nonzero total benefit. The example was artificial because there was only one origin and one destination and the routes had no common links, but the same principle may be even more convincing for multiple origin-destination flows.

Suppose that we initially had a rectangular grid of roads such as those described in section 7.6. The trip table, cost functions, and the network are invariant to translations in either the N-S or E-W directions; consequently, the link flows are also invariant to translations. We now make an improvement in one road throughout its entire length; for example, the W-E road with intersections along $[0, i]$, $-\infty < i < +\infty$. The cost of travel per block length for this road is changed from $c_E(\cdot)$ to $c'_E(\cdot)$, with $c'_E(f) < c_E(f)$ for all values of f. All other W-E roads remain as before with cost functions $c_E(\cdot)$.

Starting from any route assignment that existed prior to the improvement, for example, all trips traveling the W-E or E-W segment first and then the N-S or S-N segment, we reassign trips to the improved road I by allowing some trips that go from one side of I to the other to use I for the W-E segment. Thus, in figure 7.4, trips that originally went by way of a route A-B-C could be reassigned to the route AA'-B'-C. As we do this, we remove flow from W-E roads adjacent to I (for example, the links AB), thereby reducing the cost of travel on these roads while increasing the travel cost $c'_E(f'_E)$. But there are also trips between O-D pairs that are not on opposite sides of I; in particular, there are trips with destination B that previously were traveling by way of the route D-E-B of figure 7.4. If there is a decrease in congestion on routes such as A-B, the trips from D to B would prefer to go by way of the route D-A-B. This would decrease the cost

Assignments on Idealized Networks

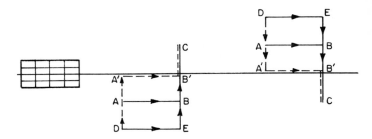

Figure 7.4
Reassignment of trips on a square
grid.

on *D-E* and induce trips with destination at *E* to use the
segment *D-E*. By such a chain reaction, the effect of the
improved road can propagate arbitrarily far in the N-S
direction.

If all trips that cross *I* were assigned a route that uses *I* for
the W-E leg, as in figure 7.4, then the flow along the link
$[0, 0]$ to $[1, 0]$ of *I* would include all trips from origins
$[i, j]$ to destinations $[i', j']$ for which $i \leq 0$ and $i' > 0$ and
either $j \leq 0, j' \geq 0$ or $j \geq 0, j' \leq 0$. Equivalently, for trips
from $[i, j]$ to $[i + k, j + l]$ for fixed k, l, it would include all
trips with $-k \leq i \leq 0$ and $-l \leq j \leq 0$ if $l \geq 0$ (or $0 \leq j \leq -l$
if $l \leq 0$); that is, $k(|l| + 1)q(k, l)$ trips. The total flow on *I*
would be

$$f_I = \sum_{k>0} \sum_{l} (|l| + 1)kq(k, l). \tag{7.13}$$

This should be compared with the original flow on any
W-E road,

$$f_E = \sum_{k>0} \sum_{l} kq(k, l). \tag{7.14}$$

The flow f_I is larger than f_E by a factor comparable with
the average N-S trip length component measured in block
lengths. Unless the improvement in *I* consists of building a
high-capacity freeway or a transit line, one would not ordinarily expect that the improved facility could carry a flow f_I.
Even if it could, it is likely that $c'_E(f_I)$ would be larger than
$c_E(f_E)$.

Suppose some fraction of the trips that cross I use route I so as to achieve a uniform flow of f'_E, $f_E < f'_E < f_I$. If all other trips were to choose any cheapest route without using I, they would distribute themselves over the network in such a way that all other W-E roads would have the same travel cost; for if any road were cheaper, it would immediately attract flow from all trips crossing it.

If, before the improvement, all trips traveled first along the W-E leg of the trip and then the N-S leg, they would now travel the W-E leg along some road nearer to I. In the extreme case $f'_E = f_I$, we would assign all trips from $[i, j]$ to $[i', j']$ to use the W-E road through $[i, j]$ or $[i', j']$, whichever is closer to I.

Because this rerouting has no effect on the total N-S or S-N flows, f_N and f_S remain unchanged. For the W-E traffic, the extra flow $f'_E - f_E$ that is assigned to the new route must be subtracted from other W-E roads. Because the flows are all equal on the latter, very little flow is taken from any one road (or if it is, it is replaced by flow from other roads). This flow, therefore, also remains essentially unchanged at the original value f_E.

If all trips were allowed to choose the cheapest route, the flow f'_E would automatically adjust so that

$$c'_E(f'_E) = c_E(f_E), \tag{7.15}$$

provided that the resulting f'_E is less than f_I. Again it appears that no one gains individually; yet there is a positive total benefit. If any flow ε_k is removed from $[i, k]$ to $[i + 1, k]$ of the W-E link, the total cost for the trips that remain on this link is reduced by an amount

$$f_E \varepsilon_k dc_E(f_E)/df_E.$$

The total benefit in a N-S strip from $[i, k]$ to $[i + 1, k]$, $-\infty < k < \infty$, is, therefore,

$$\left[\sum_{k=-\infty}^{\infty} \varepsilon_k \right] f_E dc_E(f_E)/df_E.$$

Even though the ε_k are all individually very small, the total flow $\Sigma \varepsilon_k$ that is removed from this strip must be equal to the flow $f'_E - f_E$ that is added to the link ($[i, 0]$, $[i + 1, 0]$) of I. Thus the benefit per block length of improved road is

$$-\Delta T = (f'_E - f_E) f_E dc_E(f_E)/df_E. \tag{7.16}$$

If (7.15) is true, then the (finite number of) trips that shift to any link of I receive essentially zero benefit; it is the (infinite number of) trips that do not shift that receive a nonzero total benefit. If, on the other hand, one should assign any flow f'_E to I, such that $c'_E(f'_E) < c_E(f_E)$, the benefit to those trips that do not shift is still given by (7.16), but the trips assigned to I receive a benefit per block length of

$$f'_E [c_E(f_E) - c'_E(f'_E)]. \tag{7.17}$$

The total benefit is the sum of (7.16) and (7.17),

$$-\Delta T = f'_E[-c'_E(f'_E) + c_E(f_E) + f_E dc_E(f_E)/df_E] \\ - f_E^2 dc_E(f_E)/df_E. \tag{7.18}$$

The choice of f'_E that maximizes (7.18) can be determined by setting the derivative of ΔT with respect to $f'_E = 0$. This yields the obvious condition that the marginal costs be equal on the two routes,

$$c'^*_E(f'_E) = c^*_E(f_E) \tag{7.19}$$

and

$$(-\Delta T)_{max} = f'^2_E \frac{dc'_E(f'_E)}{df'_E} - f_E^2 \frac{dc_E(f_E)}{df_E}. \tag{7.20}$$

That one can achieve a benefit (7.20) instead of (7.16) by choosing an f'_E to satisfy (7.19) instead of (7.15) is somewhat academic. In addition to the usual controversies associated with the fact that travelers do not really agree on the cost functions and that regulation of the flow is difficult to implement, there is a further complication that the benefit (7.20) applies only if the flow f'_E comes from the trips that cross I.

Typically one expects the flow f'_E satisfying (7.19) to be less than the flow satisfying (7.15); so one must artificially restrict the flow on I to achieve (7.19) and $c'_E(f'_E) < c_E(f_E)$. Because trips crossing I benefit more from using I than trips that must backtrack to reach I, it is necessary, in order to achieve the benefit (7.20), to exclude from f'_E any trips not crossing I. There is, however, no practical mechanism for regulating the entrance to I that can distinguish travelers who will backtrack from those who will not. Although (7.16) was also derived under the hypothesis that only trips crossing I will use it, if (7.15) is true for $f'_E < f_I$, the equilibrium assignment in which each traveler chooses a cheapest path will automatically make it unprofitable for any other trips to use I; all trips will go by way of a shortest distance route.

One should notice that the benefit (7.16) depends on the amount of flow f'_E attracted to I for a cost on I of $c_E(f_E)$ and on the congestion on the network; but the benefit is otherwise independent of the properties of I. If there is no congestion on the network, $dc_E(f_E)/df_E = 0$, there is no benefit. Although the benefit depends on the flow f_E on the network, it is otherwise independent of the O-D table; that is, the trip length distribution.

The benefit (7.16) is proportional to $f'_E - f_E$. If one were to improve two roads I_1 and I_2, and there were enough trips crossing I_1 and/or I_2 to establish an equilibrium with travel cost on both I_1 and I_2 of $c_E(f_E)$, the benefit would simply be the sum of the benefits for each, independent of the spacing between I_1 and I_2. In particular, if I_1 and I_2 were to virtually coincide, they would attract a combined flow of $2(f'_E - f_E)$ at a benefit of twice that of a single road.

One may question whether very small cost reductions (ε_k) applied to a large number of different trips are worth anything compared with the corresponding sum of benefits applied to just one or a few trips. One could conclude that there is really no benefit at all from an improvement in only one road. The main point of this analysis is that effects of an improvement in only one road propagate infinitely far on an infinite grid. Relative to an infinite grid, however, an improvement in only one road represents an infinitesimal investment per unit area. If one expects to improve the cost of travel over the entire grid (or any part of it), one must make some nonzero increase in the average capacity per unit area, but the benefit from such an increase in capacity is still not sensitive to the detailed location of the improvements.

These conclusions were based on the assumption that the trips crossing the new route, or routes, could generate enough traffic along the route to make the travel cost per unit distance on it equal to that on the adjacent streets. If the improvement consisted of building a freeway (or transit line) of high speed and large capacity, the freeway could attract all the flow f_I that crossed it and then start to draw additional trips from neighboring streets. The latter trips, however, must increase their travel distance (backtrack) in order to use the freeway. A similar result will obtain if one builds many new facilities at such a spacing that some trips cross several new facilities. The flow attracted by any one facility will then be less than f_I because these longer trips would be shared among several routes. These more general problems can also be analyzed, but the evaluation of the benefits is more complicated. The benefit now depends on the trip length distribution because the trips with longer eastbound leg can afford to backtrack longer distances in the N-S direction than short trips to take advantage of the lower cost on the new routes.

To illustrate a few interesting issues concerning backtracking to parallel freeways (or transit lines), we will consider a problem of determining an optimal spacing between freeways, but only to serve trips having a very artificial trip length distribution. The problem set at the end of the chapter gives other rather artificial examples with backtracking. For more general illustrations with arbitrary trip length distributions (but rather tedious algebra) involving similar issues see references 3, 4, and 5.

7.8 Parallel Freeways Suppose one were to superimpose a family of parallel freeways running E-W at a spacing S on a fine (spacing small compared with S) rectangular grid of streets running N-S and E-W. The flows on the grid are assumed to be sufficiently low that we can disregard congestion effects and consider the cost per unit distance of travel to be c in either direction. The freeway interchanges are sufficiently close together (compared with the trip length) that we can disregard longitudinal access costs. The cost per unit distance of travel on a freeway, however, is considered to be dependent on flow and represented by $c'(f)$, with $c'(0) < c$.

Trip origins are uniformly distributed over the plane with ρ trips per unit area, and each trip from a point $[x_1, x_2]$ is assumed (unrealistically) to have a destination $[x_1 + L, x_2]$ traveling only in the eastward direction a distance L. Trips are assigned according to either the user-optimal route or the system-optimal route.

The spacing S is unspecified. If it costs C units of money per unit distance and unit time to own and operate the freeways (C is measured in the same units as $f c'(f)$), we will choose S so as to minimize the sum of investment and travel costs per unit area.

For a trip with origin (and destination) a distance x_2 from a freeway, $|x_2| < S/2$, the cost of travel by way of the freeway is $2|x_2|c + Lc'(f)$, whereas the cost by way of the E-W streets is cL. If it is cheaper to travel by way of the freeway for one value of $|x_2|$, it is cheaper also for any smaller value of $|x_2|$. Thus there is a maximum value of $|x_2|$, $S_0/2$, where

$$S_0 = [1 - c'(f)/c] L, \tag{7.21}$$

such that if $S_0 < S$, trips with $S_0/2 < |x_2| < S/2$ will prefer to travel by way of the streets and those with $0 < |x_2| < S_0/2$ will prefer the freeway. The flow f passing any point on the freeway will include all trips originating within a rectangle having E-W sides of length L and N-S sides of length S_0; that is, $f = S_0 L \rho$.

Correspondingly, if trips are assigned so as to minimize the total transportation cost, there is an S_0^*, where

$$S_0^* = [1 - c'^*(f)/c] L < S_0, \tag{7.22}$$

such that for $|x_2| < S_0^*/2$, the marginal cost of travel by way of the freeway is less than by way of the E-W streets. If $S_0^* < S$, the system optimal flow on the freeway would be $f = S_0^* L \rho$.

From graphs of $c'(f)$ and $c'^*(f)$ one can easily determine the values of S_0 and S_0^* and the corresponding values of f that satisfy (7.21) and (7.22), but it is immediately clear that the optimal choice of S, S^*, must be less than S_0^* (and S_0) if it is profitable to have any freeways at all; that is, if $S^* < \infty$. One argument is that if S were larger than S_0^*, there would be an E-W strip of width $S - S_0^*$ between the freeways from which the travelers should be assigned to the street route to achieve the system optimal, and if S were larger than S_0, there would even be a strip of width $S - S_0$ in which the travelers themselves would prefer the E-W street route. The benefits from the freeways would not change, however, if n of the freeways were displaced in the N-S direction until their shed boundaries just touched (that is, for the system-optimal assignment, the spacings were reduced to S_0^*), leaving a strip of width $n(S - S_0^*)$ that would not be served by a freeway. But if it were worthwhile to build freeways to serve the travelers in the strip of width nS_0^*, it must also be profitable to build freeways to serve those in the strip of width $n(S - S_0^*)$, which, for sufficiently large n, will be at least S_0^* wide. Thus, the proposed spacing is not optimal.

For this artificial trip distribution, we see that by choosing $S < S_0^*$ we dispose of any question as to whether trips should be assigned according to a user-optimal or a system-optimal. All trips that prefer the freeway benefit enough to justify the congestion they cost others. One need not charge tolls in order to induce travelers to choose an optimal route. Indeed, for this trip distribution, no one will use the E-W streets, consequently, there is no obvious reason why they should be there. (This issue will be explored further in chapter 9.)

Because the S_0^* depends on L, one cannot generally determine a spacing S so as to induce a system-optimal assignment simultaneously for a distribution of L values. This exercise does, however, further illustrate (and exaggerate) the point made in section 7.3: There are parameters associated with the network geometry that can be selected in such a way as to restrict the opportunities for travelers to select routes that are costly to the system.

To evaluate the optimal S, one can minimize either the average cost per trip or the cost per unit area, or maximize the benefit per unit area. For $S < S_0^*$ all trips use the freeway and backtrack an average distance $S/4$ to and from the freeway (total $S/2$). The average benefit per trip using the freeway is, therefore,

$$L[c - c'(f)] - cS/2,$$

and the average travel benefit per unit area is this multiplied by the number of trips per unit area, ρ. The length of freeway per unit area is $1/S$, and the cost per unit area is C/S. The total net benefit per unit area for $S < S_0^*$ is, therefore,

$$\rho L[c - c'(f)] - c\rho S/2 - C/S, \tag{7.23}$$

in which the flow f has the value

$$f = SL\rho. \tag{7.24}$$

To determine the optimal S and S^*, we can set the derivative of (7.23) with respect to $S = 0$ and obtain the equation

$$-\rho^2 L^2 \frac{dc'(S^*L\rho)}{d(S^*L\rho)} - \frac{c\rho}{2} + \frac{C}{S^{*2}} = 0. \tag{7.25}$$

From a graph of $dc'(f)/df$, one can easily locate the solution S^* of (7.25), but these formulas are valid only if $S < S_0^*$.

If we use (7.25) to eliminate C from (7.23), we see that the maximum benefit at $S = S^*$ has the value

$$\rho c\{L[1 - c'^*(f)/c] - S^*\}.$$

If we compare this with (7.22), we see that the net benefit per unit area is positive only if the solution S^* of (7.25) satisfies the condition $S^* < S_0$. Conversely, if the net benefit is positive for some S so that it is worthwhile to build freeways, the value of S^* will be less than S_0^* (as predicted earlier by a somewhat different argument).

We can also write (7.25) in the form

$$S^{*2} = \frac{2C/c\rho}{1 + \dfrac{2\rho L^2}{c}\dfrac{dc'(S^*L\rho)}{d(S^*L\rho)}} \le \frac{2C}{c\rho}. \tag{7.26}$$

If the freeways are not too congested, one can disregard the term containing $dc'(f)/df$ (which depends on S^*) and obtain the simple approximation

$$S^* \simeq [2C/c\rho]^{1/2}. \tag{7.27}$$

The interesting feature of this approximation is the parameters that are missing from the formula. It does not contain L, which would imply that it is valid even for certain distributions of trip lengths (provided that the condition $S^* < S_0^*$ is valid for all L values). Neither does it contain $c'(f)$, which means that the optimal spacing is (nearly) independent of the properties of the freeway, or whatever, provided that it was worth building.

This last property may seem counterintuitive, but it is not very sensitive to the artificial features of this illustration. Basically, the choice of an optimal spacing involves a balance between construction costs, represented in (7.26) by C (the smaller S, the higher the cost), and access costs, represented in (7.27) by $c\rho$ (the larger S, the further one must travel to reach a freeway). The cost of the E-W leg of a trip obviously depends on the facility, the $c'(f)$, but this cost does not depend on the spacing S except that the value of S affects the flow f, which, in turn, influences the value of $c'(f)$. The latter effect is represented in (7.26) by the term in the denominator, which is neglected in (7.27).

7.9
Further Extensions
The analysis of the last section exploited the E-W translational symmetry of an idealized traffic assignment problem, but the inclusion of one or more improved E-W routes destroyed the N-S translational symmetry used in section 7.5. More generally, one could consider a rectangular grid network for which the roads and trip distribution are invariant to translations in the E-W direction but have arbitrary variation in the N-S direction; that is, all E-W

roads may be different (but homogeneous along their path), and trips from different E-W roads may have different generation rates and trip length distribution, but all N-S roads are identical and equally spaced.

By exploiting the E-W symmetry (which implies identical flow patterns on all N-S roads), this two-dimensional assignment problem can, in effect, be reduced to a one-dimensional problem in which the assigment is specified when each trip with an origin on one E-W road and destination on another E-W road selects the E-W road along which it will make the E-W leg of its trip. (This more general problem has been analyzed in considerable detail in reference 4.) Although this is still a rather special class of problems, it includes an enormous variety of examples that can be described completely.

It is also possible to analyze in considerable detail the class of traffic assignment problems on a ring-radial–type network that is invariant to rotations through an angle $2\pi/n$, n being an integer; that is, networks with identical radial routes (regularly spaced) but arbitrary homogeneous ring roads and a trip distribution with arbitrary dependence on the radial coordinates. The rotational symmetry, in effect, again reduces a two-dimensional problem to a one-dimensional problem; each trip between an origin on one ring road and destination on another must select the ring road along which it will make the circular leg of its trip. (Some of these problems are analyzed in reference 6. A complete transportation planning study based on a rotationally symmetric city is discussed in reference 7.) There are certain qualitative aspects of these special problems that are relevant to the understanding of the traffic assignment on general networks according to schemes described in chapter 6.

Most real transportation networks are similar to a rectangular or ring-radial network, at least in the sense that very few trips can go directly from their origin to destination along a single road. Each leg of a typical trip intersects many other roads that are potential candidates for a second leg of the trip. Even though there might be a unique cheapest route between the origin and destination, there is typically an enormous number of other possible routes with comparable length.

If any road is clearly superior to others running (nearly) parallel with it under conditions of light traffic, it will attract most of the traffic that could use (a potentially very large number of) adjacent roads. Most systems are (and should be) designed so that the most popular facilities are congested at least during the rush hour. At some moderate level of traffic the flow on these facilities must spill over onto the other adjacent roads. For a ring-radial network, for example, ring roads near the center should become congested rather easily, but traffic that might use inner ring roads at light traffic levels will use ring roads farther out as the traffic level increases.

At sufficiently high traffic levels (in all directions) any nearly rectangular cell of a network will have the property that the cost of travel between any two diagonal corners of the cell is independent of the path (clockwise or counter-clockwise). For any of the networks with symmetry in one direction, the cost of travel already is the same on one pair of opposite sides of the cell so that this condition forces the travel costs to be equal on the other two opposite sides also.

For the rectangular grid this means that the cost of travel per unit length is the same on adjacent parallel roads, even though the roads may be quite different. For the ring-radial network it means that the cost of travel per unit angle is the same on adjacent ring roads; that is, the cost per unit distance on a ring at radius r is proportional to $1/r$.

What is meant by "sufficiently high traffic levels" varies from place to place in the network so that for any given trip distribution there will be some regions of the network where this condition is true and other regions where it is not. A region with high traffic levels generally will include all roads enclosed by some simple closed curves in the two-dimensional plane because trips have great flexibility in switching between neighboring roads and would not allow any road in the interior of a region to retain a superiority. As the traffic level increases, this congested region of the network expands.

In these symmetric problems, most of the mathematical

complexity is associated with defining the boundaries of the regions of congestion and determining the travel costs in the regions where adjacent roads do not have equal travel costs. The latter is further complicated by the fact that some trips will backtrack to take advantage of any cheaper routes (long trips, particularly, can divert to go around the congested region).

If, in some specified region, one knew that the travel costs would be equal on parallel links, one would not need very much information to specify the actual values of the flows and travel costs. Certainly one would not need to know the complete trip distribution or the routing of trips. If, for example, one knew the cost per unit distance of travel c along a family of parallel roads, one could deduce from the $c_j(f_j)$ curves for the individual roads what value of f_j would give $c_j(f_j) = c$. By adding the flows f_j over a family of parallel roads, one would have a relation between the cost c and the flux. One could also replace a family of parallel roads by a single "effective road" with an appropriate $c(f)$, $f = \Sigma f_j$. Conversely, if one knew the vehicle-miles of travel per unit area in various directions, one could deduce the cost of travel. Note that the construction of this $c(f)$ is essentially that described by figure 6.2.

One possible way to give a crude representation of a congested region would be to replace the existing network on this region by a coarse square grid, with each new road representing a family of parallel roads of the original network in the sense described earlier. The conventional planning procedures could be used to estimate the flows and travel costs on this coarse grid. Having established the cost per unit distance of travel in various directions, one could then return to the original network and evaluate the flows on individual roads (if relevant).

In any case, anything one can learn about traffic behavior from the study of simple idealized models should be exploited, somehow, to simplify or improve the accuracy of conventional procedures.

Problems 1

A very wide freeway of width w carries flow in two directions. By moving a dividing strip, it is possible to partition the width w into

$$w = w_E + w_W,$$

w_E for eastbound traffic, w_W for westbound traffic, for any value of w_E, $0 < w_E < w$. For an eastbound flow f_E on a road of width w_E, the cost per trip-mile of eastbound travel is $c_E(f_E; w_E)$. Similarly the cost per trip-mile of westbound travel is $c_W(f_W; w_W)$. These costs have the form

$$c_E(f_E; w_E) = c(f_E/w_E), \qquad c_W(f_W; w_W) = c(f_W/w_W),$$

in which $c(f)$ and its derivative are monotone increasing functions of f. How should one choose w_1 and w_2 so as to minimize the total travel cost?

2

Two transportation routes run parallel to each other. At any point x along the route, a traveler may switch from one route to the other (in either direction) at a cost of δ. All trips start and end on route 2, so that a traveler must pay at least 2δ to use route 1. The cost per unit distance of travel on route 2 is c_2, independent of position and the flow. The cost per unit distance of travel on route 1 at coordinate x is $c_1(f_1(x))$ if the flow at x is $f_1(x)$. The function $c_1(f)$ is monotone increasing, and $c_1(0) < c_2$.

There is a given density of trips from x to y such that

$$\rho'(x, y)\, dx\, dy = \text{number of trips per unit time with origin}$$
$$\text{between } x \text{ and } x + dx, \text{ destination between}$$
$$y \text{ and } y + dy.$$

This density is assumed to be a function of $y - x$ only and vanishes for $y < x$ (all trips are to the right); that is,

$$\rho'(x, y) = \begin{cases} g(y - x) & y > x, \\ 0 & y < x, \end{cases}$$

for some function $g(z) \geq 0$. If every traveler chooses the cheapest route to himself, how would one evaluate $f_1(x)$?

3

A freeway (or transit line) with velocity of travel V has limited access, with interchanges (stations) at spacing D. Another slow-speed access road (bus line) with velocity v runs parallel with it and is fed by a dense family of perpendicular roads so as to create, in effect, a uniform density ρ' of trip origins along the access road; that is, the number of trips "originating" between x and $x + dx$ is $\rho' \, dx$. The trips are all sufficiently long (at least $2D$) that they will use the freeway (transit line) and are equally likely to travel east or west on it. The trip length distribution guarantees that the effective destinations are also uniformly distributed along the access road. Each trip chooses a path of minimum trip time from origin to destination.

Determine

a. the shed boundary between trip origin (destination) points that will enter (leave) the freeway at $x = 0$ or $x = D$ for eastbound trips (and westbound trips); and

b. the flow $f(x)$ on the access road for $0 < x < D$.

4

On a fine rectangular road grid, with coordinates $[x_1, x_2]$ along the grid directions, there are ρ trip origins per unit area, which are independent of $[x_1, x_2]$. Each trip from $[x_1, x_2]$ has destination $[x_1 + L, x_2]$. The cost of travel on the grid is c per unit travel distance, which is independent of the grid flow (which is assumed to be low), and the same for travel along any grid direction. A freeway is built along the x_1-axis ($x_2 = 0$). The travel cost per unit distance is $c'(f)$ for a flow f on the freeway; $c'(f)$ is a monotone increasing function, and $c'(0) < c$. From a graph of the function $c'(f)$, determine the flow on the freeway under each of the two optimal assignment principles. Show how f varies with c, ρ, and L for given $c'(\cdot)$.

5

A network of roads consists of a family of closely spaced roads running E-W and another family of roads running N-S at spacing D. Trip origins and destinations lie only on the E-W streets. The number of trips per unit time with an origin coordinate between x and $x + dx$ on street i and a destination coordinate between x' and $x' + dx'$ on street j is

$$\rho(|x' - x|, |i - j|) \, dx \, dx'.$$

(The O-D table has translation and reflection symmetry in both x and i coordinates.)

All routes have a flow-dependent travel cost, $c_E(f_E(z))$ per unit distance on any E-W street where the flow is $f_E(z)$ at location z and $c_N(f_N)$ per unit distance on any N-S street carrying a flow f_N. The functions $c_E(\cdot)$ and $c_N(\cdot)$ are independent of location or direction of travel; they are monotone increasing and convex.

Find the flow $f_E(z)$ on the ith street at a point midway between the N-S streets if

a. each trip chooses its cheapest route; or

b. trips are assigned to minimize total travel cost.

References

1

Braess, D. "Über ein Paradoxen der Verkehrsplanung," *Unternehmensforschung* 12 (1968): 258–268.

2

Murchland, J. D. "Braess's Paradox of Traffic Flow," *Transportation Research* 4 (1970): 391–394.

3

Allen, B. L., and Newell, G. F. "Some Issues Relating to Metering or Closing of Freeway Ramps. Part I: Control of a Single Ramp," *Transportation Science* 10 (1976): 227–242; and "Part II: Translationally Symmetric Corridor," *ibid.*, pp. 243–268.

4

Jeevanantham, I. "Flow Dependent Traffic Assignment on a Rectangular Grid Network with Translational Symmetry in One Direction," Ph.D. Thesis, University of California, Berkeley, 1972.

5

Fawaz, W. Y., and Newell, G. F. "Optimal Spacings for a Rectangular Grid Transportation Network," *Transportation Research* 10 (1976): 111–129.

6

Lam, T. N., and Newell, G. F. "Flow Dependent Traffic Assignment on a Circular City," *Transportation Science* 1 (1967): 318–361.

7

Tanner, J. C. "A Strategic Model for Urban Transport Planning," *Proceedings of the Fifth International Symposium on the Theory of Traffic Flow and Transportation.* American Elsevier, 1972.

8 NONIDENTICAL TRAVELERS

It has been postulated that there exists a uniquely defined cost function $c_k(\cdot)$ for travel on route k. This cost, which combined actual cost and cost equivalents of time, comfort, and so forth, was generally considered to be a function of the link flows on all links along the route or of all the route flows. The implication of this postulate, however, was that although people certainly do not all agree on the relative merits of various routes, the flows or other quantities deduced from such a postulate would be approximately the same as those from a more realistic hypothesis: There are many functions $c_k(\cdot)$ for each kth route, possibly a different one for each traveler, and each traveler chooses what he considers to be the cheapest route.

It is easy to construct mathematical models that include more and more variables and are, therefore, presumably more realistic, but models currently in use already contain more input variables that one can afford to measure. Rather than adding more variables into the models, one would like to reduce the number but make sure that the ones retained are the most important ones for the intended purposes, such as variables to evaluate flows, or changes in flows, resulting from changes in the network or demands. These variables should be easy to measure or have predictable future values.

Even though travelers do not agree on the travel costs of various routes, there is no practical way to make direct measurements of the probability distributions for the costs that travelers would assign to various routes. The best that can be done is to propose some plausible (but rather arbitrary) parametric family of stochastic models for the joint probabilities of various costs, preferably involving only a few parameters, and then select values of the parameters so that the model gives a best fit to some suitable observations of phenomena that are sensitive to travelers choices.

It is not our intention here to review the very extensive literature on the application of such models because this has been done by many authors. (See, for example, reference 1 or references 3 and 4 of chapter 1 and reference 10 of chapter 6.) Our purpose here is to discuss some of the possible consequences of the fact that there are different types of travelers with different cost evaluations, particularly how this affects the pattern of flow over a network.

The simplest type of assignment postulate considered in chapter 6 was that travelers always choose the cheapest route; they all agree on the costs of the routes and, therefore, which routes are the cheapest. If, in addition, we assume that the costs do not depend on the flow, we are led to an all-or-nothing assignment. In some cases the flows evaluated according to this postulate give absurd results; some routes have flows exceeding the capacity of the route while other routes have no flow. This is particularly true for networks with only a few origins and destinations or those having corridors with heavy flows along certain parallel routes.

On the other hand, if there is a near continuum of origins and destinations, as on the square grid of roads considered in chapter 7, it does not seem to make much difference what assignment principle is used; any number of possible schemes would lead to the same flows, provided that there exists some guarantee that streets become equally congested.

The failure of the all-or-nothing assignment to give feasible flows on some routes can be overcome by the introduction of capacity constraints or flow-dependent costs. This was done in chapter 6 and is done in the large transportation studies, at least for most routes that become congested. However, this is not the only way of avoiding such difficulties, nor is it the only method used in transportation planning.

We have not explicitly specified that all routes were to be interpreted as highways; some could have been other modes of travel, such as rapid transit lines or airlines. One cost $c_k(\cdot)$ could have been associated with a rapid transit line, another with the cost of a highway. We would have concluded from our postulates that, for the appropriate choice of the $c_k(\cdot)$, the flows on alternate routes between the same origin and destination would divide so that the costs are equal if both routes are used. The formal theory does not recognize any distinction between a traveler choosing between two highway routes or between routes involving different modes; in principle, such possibilities are incorporated into the form of the c_k.

Transportation planners do not usually treat the assignment of traffic to routes of widely dissimilar properties by the methods described here. Routes of different types are usually treated separately. Within the network of similar routes, traffic may be assigned according to the principle of minimum cost, but the choice between types is treated separately. The latter is described as "modal split," but it is a matter of judgment as to what types of routes are to be considered in the same mode. From the mathematical point of view, we simply make some classifications of routes and use a different assignment rule for the choice between routes of different type and the same type. In a more abstract sense, however, this means that on the complete transportation network one is employing a more general type of assignment procedure than that of minimum cost.

Most theorists would probably agree that the correct conceptual theory (impossible to apply) is that travelers should be separated into various types according to income, car ownership, work habits, personality, type of vehicle, and so forth. Each person of a particular type l has his own set of costs c_k^l for the kth route. The c_k^l may, furthermore, be functions of not only the total flow on the links of the kth route but its composition as to type. For example, some types may represent trucks or buses that are on the same highway, and a cost preference by type l may depend on the fraction of trucks or buses on the kth route.

That planners classify trips into "modes" is a recognition that, on the one hand, differences in the preferences of travelers are more important for the choice between certain routes than others. On the other hand, one cannot gather the necessary data nor do the calculations implied by a theory in which differences in preferences are recognized for all choices.

In the traffic assignment procedure described in chapter 6, the partition of the flow between two routes is attributed to a desire of travelers to equalize congestion in some sense. In the modal split, the division of the flow between two routes is attributed almost entirely to differences in the preference of travelers (and has little to do with congestion). For example, a significant fraction of bus riders have no car.

8.2
To Go or
Not to Go

To illustrate different approaches to the traffic assignment or modal split problem, we will consider the simplest type of network, one origin O, one destination D, but two routes 1 and 2 (possibly two different modes, two means of travel on the same facility, or two physically separated routes). There is a reservoir of potential travelers who desire to travel from O to D.

In our previous analysis of the assignment problem, we postulated that there was a fixed flow q that wished to travel from O to D. Economists are likely to consider a more general problem in which the flow (demand) is a function of the cost (in some sense); that is, there is an "elasticity of demand." We could incorporate an elasticity of demand, perhaps in a more realistic way than is usually done, by imagining a hypothetical third route from O to D, which is not actually a route but a choice. The choice is: Do not go. Each traveler now makes a choice among 1, 2, or 3.

If there were three choices and all travelers were identical, then all travelers would associate the same cost c_3 with the third route. This can be interpreted as the profit for the trip because if everyone takes the cheapest (most profitable) route, they would choose 3 (not to go) if the travel cost to go was greater than the profit. We will assume, as in chapter 6, that the costs $c_1(\cdot)$ and $c_2(\cdot)$ are functions of only one variable each, the flows f_1 and f_2 on these two routes, respectively. The cost c_3, however, differs from $c_1(\cdot)$ and $c_2(\cdot)$ in that the cost of this choice by one person is not likely to depend on how many other trips f_3 choose it. In effect, the choice 3 behaves like a third route with no congestion (or economies of scale).

If all travelers agree on these costs, we can determine how a flow q would be distributed among the three choices in the

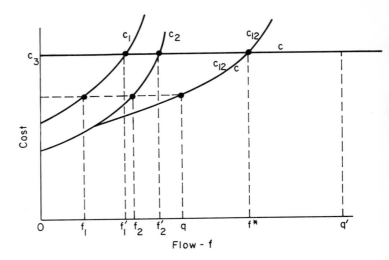

Figure 8.1
Assignment of travelers to two
routes plus a choice of no trip.

same way as in figure 4.2, except that the cost $c_3(f_3)$ is now
independent of f_3, as illustrated in figure 8.1. The horizontal
addition of the curves $c_1^{-1}(c)$ and $c_2^{-1}(c)$ yields the curve
labeled c_{12}, which represents the cost of travel $c_{12}(f)$ from
O to D for any combined flow $f = f_1 + f_2$. If we interpret
$c_3(f_3)$ to include the vertical jump from O to c_3 at $f_3 = 0$,
then the horizontal addition of $c_3^{-1}(c)$ to the other two
would give the curve labeled c, which follows the curve
c_{12} for $c_{12}(f) < c_3$ or f less than some flow f^*, $c_{12}(f^*) = c_3$,
but for $f > f^*$, $c(q) = c_3$.

This figure shows the obvious partition of any flow q, as
indicated by the broken lines. If one draws a vertical line to
the curve $c(q)$ and then a horizontal line at height $c(q)$, the
horizontal line will cut the curves $c_1(f_1)$, $c_2(f_2)$, $c_3(f_3)$ at the
flows f_1, f_2, f_3 corresponding to the specified value of q. For
$q < f^*$, this construction gives $f_3 = 0$ (everyone goes). For
$q' > f_3$, only a flow f^* uses routes 1 or 2; $q' - f^*$ uses 3
(does not go). The flows f_1' and f_2' of figure 8.1 are, of course,
independent of q'.

In more conventional economic terminology, the curve
$c_{12}(f)$ would be interpreted as a "supply curve." The

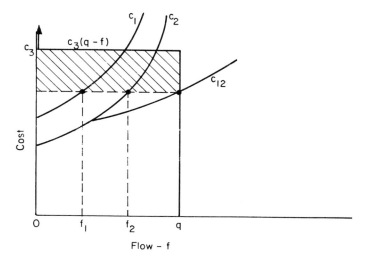

Figure 8.2
Supply-demand equilibrium
between two routes and a single
cost for no trips.

equilibrium between $c_{12}(f)$ and $c_3(f_3)$ would be evaluated by
drawing a figure analogous to figure 6.3; that is, one would
draw a graph of $c_3(q - f)$ versus f, as in figure 8.2, and call it
a "demand curve." If we think of this curve with c as the
independent variable, it would be interpreted as a graph of
the number of trips that would go at any value of c rather
than the number that would not go.

For $c < c_3$, the demand is q, independent of c. We would
say that the demand is inelastic with respect to the price c.
At $c = c_3$ or $0 < f < q$, one might say that the demand was
perfectly elastic; the price is independent of the flow. For
$c > c_3$, one can interpret the demand curve to give $f = 0$;
no one goes if the price is greater than c_3.

If everyone makes the cheapest choice, as in figure 8.1, the
equilibrium flows would be determined by the intersection of
the supply curve c_{12} and the demand curve. The component
flows f_1 and f_2 on routes 1 and 2 would be determined by the
intersections of a horizontal line with $c_1(f_1)$ and $c_2(f_2)$, as
shown in figure 8.2 by the broken lines.

If we interpret $c_3 - c_{12}(f)$ as the net profit for each trip when there is a combined flow of f on routes 1 and 2 and a cost of travel $c_{12}(f)$ (the advantage of choosing 1 or 2 rather than 3), then, for $c_{12}(q) < c_3$, the equilibrium assignment would yield a total net profit from all trips of $q[c_3 - c_{12}(q)]$. This is represented geometrically in figure 7.2 by the shaded area of the rectangle of base q and height $c_3 - c_{12}(q)$. If, however, $c_{12}(q) > c_3$, the equilibrium would yield $c_{12}(f) = c_3$ at a flow f^*—zero profit individually or collectively.

This illustrates an extreme example of the differences between a social-optimal and a user-optimal. If $q > f^*$ and if there were a flow f on routes 1 and 2 such that $f < f^*$— that is, $c_{12}(f) < c_3$—then it would be profitable for some of the remaining $q - f$ trips also to travel. If everyone makes the best personal choice, then f would increase until $f = f^*$ and no one would gain anything.

If one could artificially restrict the flow to a value $f < f^*$, there would be a positive profit. If the trips in this restricted flow f could choose the cheaper of routes 1 or 2, the net total profit would be

$$f[c_3 - c_{12}(f)]. \tag{8.1}$$

This would be a maximum for flow f, chosen so that

$$\frac{d \cdot f c_{12}(f)}{df} \equiv c_{12}^*(f) = c_3; \tag{8.2}$$

that is, the marginal cost on the route pair 1 and 2 would be equal to c_3. Note that because the cost of the choice c_3 does not depend on the flow f_3, the marginal cost of the choice 3 is the same as the individual cost.

There is a slight complication here in that (8.2) might have more than one solution and (8.1) more than one local maximum. If $c_1(0) > c_2(0)$, as in figure 8.1, the derivative of $c_{12}(f)$ has a discontinuity at $f = f_0$ for which $c_2(f_0) = c_1(0)$, where travelers start to use route 1. Therefore the marginal cost $c_{12}^*(f)$ also has a discontinuity at f_0. It is possible, for $f < f_0$, that $c_{12}^* = c_2^*(f)$ (which is assumed to be an increasing function of f) reaches the value c_3 already for $f < f_0$ and gives one local maximum for the profit. At $f = f_0$, however, $c_{12}^*(f)$ drops, possibly from the value above

c_3 for f just less than f_0, to a value below c_3 for f just larger than f_0. For $f > f_0$, $c^*_{12}(f)$ should increase again and could cross c_3 a second time, giving a second local maximum for the profit. There is no way of automatically determining which of the local maxima will be the true maximum; one must evaluate the profit at each and compare them.

More generally, if the flows f_1 and f_2 on routes 1 and 2 could be assigned separately (without the restriction that each traveler can choose the cheaper route), the total profit would be

$$(f_1 + f_2)c_3 - f_1 c_1(f_1) - f_2 c_2(f_2). \tag{8.3}$$

This would be maximized by setting the partial derivatives with respect to both f_1 and f_2 equal to zero; that is, the f_1 and f_2 would satisfy the conditions

$$c^*_1(f_1) = c^*_2(f_2) = c_3. \tag{8.4}$$

The marginal costs should be equal on all three choices. If the $c^*_1(f)$ and $c^*_2(f)$ are increasing functions of f, (8.4) will have a unique solution for f_1 and f_2. This solution will, of course, give a larger net profit than (8.1).

8.3
Elastic Demand

This main weakness of this theory is that travelers will not agree on the values of $c_1(f_1)$, $c_2(f_2)$, or c_3. This is particularly true of c_3 because different travelers obviously receive different benefits from making a trip. As a first generalization, suppose we now assume that travelers agree on the $c_1(f_1)$ and $c_2(f_2)$ but receive a different benefit c_3.

To describe the distribution of benefits among different travelers, it suffices to specify a function,

$f_d(c) =$ number of trips between O and D that would receive a benefit greater than c
  $=$ number of trips that would be made if the cost were equal to c.

This is clearly a monotone decreasing function of c and, therefore, has an inverse,

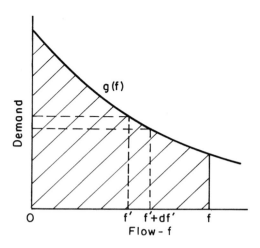

Figure 8.3
Interpretation of the demand
curve.

$$g(f) = f_a^{-1}(f) = \text{cost for which a flow } f \text{ would find it}$$
$$\text{profitable to travel from O to D.}$$

In the special case in which all travelers receive a benefit c_3, the function $g(f)$ is shown in figure 8.2 by the curve labeled $c_3(g - f)$.

To interpret these functions, it is convenient to arrange and assign the travelers in decreasing order of their benefit. If we assign a flow f to trips from O to D, it will always be to those trips with the greatest benefit; that is, the trips with benefit greater than $g(f)$. In figure 8.3, this means that the trips associated with the increment of flow from f' to $f' + df'$ are those that receive a benefit between $g(f')$ and $g(f' + df')$.

If there is a flow f', and we assign an increment of flow df', the total benefit to the trips df' is approximately $g(f')df'$. The total benefit to all trips if a flow f is assigned is the sum (integral) of all incremental benefits, which is represented by the area under the curve $g(f)$ for $0 < f' < f$:

$$\text{Total benefit for flow } f = \int_0^f g(f')df'. \tag{8.5}$$

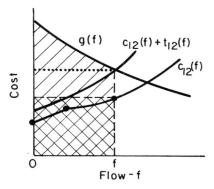

Figure 8.4
Total benefit and cost of travel.

If there is a flow f, and all trips choose the cheaper route
1 or 2, the cost per trip would be $c_{12}(f)$ for all trips. For
$c_{12}(f) \leq g(f)$, each trip would profit from travel and the net
profit for all trips would be

$$\text{Net profit} = \int_0^f g(f')df' - fc_{12}(f). \tag{8.6}$$

The second term is the area of the crosshatched rectangle in
figure 8.4; the net profit is the shaded area. This is the
generalization of (8.1).

If all travelers who find it profitable to travel are allowed
to do so, the flow f will automatically adjust so that

$$g(f) = c_{12}(f). \tag{8.7}$$

The last trip to be assigned receives essentially zero profit,
but all others receive a positive profit. The equilibrium is
established by the intersection of the curves $g(f)$ and $c_{12}(f)$
in figure 8.4. This equilibrium is the same as would exist if
there were some flow-dependent cost $c_3(f_3)$ associated with
choice 3; that is, the $g(f)$ is considered a function $c_3(f_3)$ of
the number $f_3 = q - f$ of travelers who do not go rather
than the number who do. The equilibrium would then be
defined by

$$c_3(f_3) = c_{12}(f) = c_1(f_1) = c_2(f_2); \qquad f_1 + f_2 + f_3 = q. \tag{8.8}$$

215   Nonidentical Travelers

One should notice that there is some similarity between the benefit (8.5) and the terms of T^* in chapter 6, particularly in equation (6.16). If we write (8.5) in the form

$$\int_0^f g(f')df' = \int_0^q g(f')df' - \int_f^q g(f')df'$$

$$= \int_0^q c_3(y)dy - \int_0^{f_3} c_3(y)dy,$$

the first term represents the total benefit that would be realized if all q trips were made (this is independent of f or $f_3 = q - f$). The second term, therefore, represents the cost of not traveling for the f_3 trips that do not go.

Although $c_1(f_1)$, $c_2(f_2)$, and $c_3(f_3)$ seem to enter into the equilibrium conditions in a symmetric way, the total costs associated with the flows f_1, f_2, and f_3 are interpreted quite differently. That $c_1(f_1)$ and $c_2(f_2)$ are increasing functions of f_1 and f_2 was considered to be a result of congestion on routes 1 and 2. As the flows f_1 or f_2 increase, the cost per trip increases for all f_1 or f_2; thus the total costs of these trips are $f_1 c_1(f_1)$ and $f_2 c_2(f_2)$, respectively.

There are no congestion effects for the third choice, but the travelers do not agree on the benefit. We have, however, ordered the travelers according to their benefit so that the first trips to go are those that benefit the most, or, equivalently, those who do not go are those who benefit the least. Thus, the more trips assigned to choice 3, the higher the cost to the marginal trips that are assigned not to go. The total cost to all users of choice 3 is not $f_3 c_3(f_3)$, as it would be if the cost were due to congestion. When an increment of flow df_3 is assigned to choice 3, it does not affect the cost of those previously assigned. Thus the total cost to all users of choice 3 is

$$\int_0^{f_3} c_3(y)dy, \tag{8.9}$$

the sum (integral) of all incremental costs.

To maximize the net profit of (8.6) with respect to f, we should choose an f so that

$$c_{12}^*(f) = g(f),$$
(8.10)

which is the generalization of (8.2). Because of the potential discontinuity in $c_{12}^*(f)$, there is again the possibility that (8.10) will have more than one solution and (8.6) will have more than one local maximum. Note that, in terms of the cost $c_3(f_3)$, (8.10) equates the marginal cost $c_{12}^*(f)$ to the cost $c_3(f_3)$, not to some $c_3^*(f_3)$.

If f_1 and f_2 are assigned separately, the generalization of (8.3) is

$$\text{Net profit} = \int_0^{f_1+f_2} g(f')df' - f_1 c_1(f_1) - f_2 c_2(f_2).$$
(8.11)

The maximization of this with respect to both f_1 and f_2 gives

$$c_1^*(f_1) = c_2^*(f_2) = g(f_1 + f_2)$$
(8.12)

as the generalization of (8.4).

If one could somehow regulate the flows f, or both f_1 and f_2, separately, so as to satisfy (8.10) or (8.12), there would be travelers who could profit individually from use of the facilities but who are denied use of them; specifically, anyone who receives a benefit greater than $c_{12}(f)$, $c_1(f_1)$, or $c_2(f_2)$ at the regulated values of f or f_1, f_2. A further complication is that it does not suffice simply to regulate the flows at f, or f_1 and f_2, to realize the maximum profit. In the maximization of (8.12), it was specifically assumed that travelers would be assigned in order of their benefits. Thus if the flow is regulated at a value of f or $f_1 + f_2$, all travelers who contribute to f must be those who receive a benefit in excess of $g(f)$. As a practical matter, it is, of course, rather difficult to imagine how, if one regulates the flow at a value f such that $g(f) > c_{12}(f)$, one can also make sure that those who travel are those who benefit the most; one must regulate which travelers as well as how many travelers.

In principle, one way to guarantee that only travelers who receive the most benefit will use a facility at a given flow is to charge tolls. If a single toll $t_{12}(f)$ were charged to travel from O to D when the flow is f, but each traveler could then choose the cheaper route, an equilibrium should be established at a flow such that

$$g(f) = c_{12}(f) + t_{12}(f),$$

as if the curve $c_{12}(f)$ in figure 8.4 were simply replaced by the curve $c_{12}(f) + t_{12}(f)$. At this flow, only those travelers who receive a benefit larger than $c_{12}(f) + t_{12}(f)$ will travel.

As usual, the total revenue collected is interpreted as simply a cash transfer. Although the net benefit to the users is now represented in figure 8.4 by the area above the dotted line,

$$\int_0^f g(f')df' - f[c_{12}(f) + t_{12}(f)],$$

the area $ft_{12}(f)$ between the dotted line and the broken line also represents a benefit, the total revenue collected. The net profit is, therefore, still interpreted as (8.6). Instead of trying to regulate the flow directly, the toll is adjusted to create the desired flow (and the desired choice of users). The toll that maximizes the net profit (8.6) is

$$t_{12}(f) = c_{12}^*(f) - c_{12}(f) = fdc_{12}(f)/df. \tag{8.13}$$

It is significant that the optimal toll (8.13) does not depend explicitly on the curve $g(f)$. The flow that supposedly results from this toll depends on $g(f)$, but if one knows the function $c_{12}(f)$, one can adjust the toll according to the observed f and assume that the observed f is at the intersection of $c_{12}^*(f)$ with some (possibly unknown) $g(f)$. The usual interpretation of (8.13) is that it is the cost that a "marginal user" imposes on others. Here the "marginal user" is generally considered to be the last one assigned when travelers are ordered according to their benefit, the one who receives the least benefit. Everyone, however, pays the toll and all users contribute equally to the congestion.

If one could charge separate tolls $t_1(f_1)$ and $t_2(f_2)$ on routes 1 and 2, respectively, the maximization of (8.11) would result from tolls

$$t_1(f_1) = c_1^*(f_1) - c_1(f_1) \quad \text{and} \quad t_2(f_2) = c_2^*(f_2) - c_2(f_2).$$

Again this does not depend explicitly on $g(f)$, and each toll has the interpretation as the cost that a marginal user imposes on others.

<table>
<tr><td>**8.4**
Generalization to
Networks</td><td>The theory described in section 8.3 can obviously be generalized to $n > 2$ routes between an origin and destination. It can also be formally generalized to networks.</td></tr>
</table>

Suppose that for each O-D pair i, j, there exists a demand function,

$g_{ij}(q_{ij}) = $ cost for which a flow q_{ij} would find it profitable to travel from i to j.

If travelers from i to j are ordered according to their benefit, $g_{ij}(q_{ij})$ can also be thought of as the cost that the q_{ij}th traveler between i and j associates with the choice not to go. Each traveler is considered to have a specified origin i and destination j; he simply chooses between going or not going. The total benefit of travel at a flow q_{ij} is, therefore,

$$\int_0^{q_{ij}} g_{ij}(y)dy,$$

provided that, for a flow q_{ij}, only trips with benefit larger than $g_{ij}(q_{ij})$ travel from i to j. The total benefit for all trips in the network is considered as the sum of those for all O-D pairs:

$$\text{Total benefit} = \sum_{i, j} \int_0^{q_{ij}} g_{ij}(y)dy. \tag{8.14}$$

The net benefit from travel is (8.14) less the total transportation cost. If everyone agrees on the cost of travel on the links, and the cost of a trip is the sum of the link costs, then the total transportation cost can be represented by the T defined in chapter 6.

One possible assignment scheme allows each potential traveler to maximize his own net benefit; that is, to travel by way of the cheapest route if this cost is less than the benefit to be derived from the trip. This will lead to the same user-optimal assignment discussed in chapter 6, but the flow from i to j is determined so as to satisfy the subsidiary condition that $g_{ij}(q_{ij})$ be equal to the travel cost from i to j by the cheapest route. The flow can also be determined by maximizing the total benefit less the T^* defined in equation (6.22) with respect to the q_{ij}, subject to the usual conservation equations associated with a flow q_{ij} from i to j.

If one maximizes the total net benefit, one should assign each traveler to the route with the least marginal cost and allow him to travel only if his benefit exceeds this marginal trip cost; thus the q_{ij} is chosen so that $g_{ij}(q_{ij})$ is equal to the cheapest marginal trip cost from i to j. One way to achieve this assignment is to charge tolls on each link so that the cost per trip including the toll is equal to marginal cost. The total cost of all tolls, which equals the revenue from all tolls, is, of course, not included in the total travel cost.

Although the application of this theory to networks is conceptually useful, it is of limited practical value for a number of reasons. First, the total benefit (8.14) is based on a premise that each traveler has a specific origin and destination and can only choose to go or not go. A more realistic model would presumably recognize that a traveler from i who chooses not to go to j may go elsewhere. Obviously any economic theory that includes a comparison between benefits from various destinations would have a much more complex mathematical structure. In such a theory, the number of trips q_{ij} from i to j would depend not only on the cost of travel from i to j but also on the cost of travel to other destinations.

We concluded that, for a special type of benefit function, charging tolls so that travelers must pay the marginal trip cost will lead to a maximization of the total net benefit. This conclusion does not seem to be very sensitive to the mathematical form of the benefit. It would still be true if the total benefit is the sum of the benefits to individual travelers, if each traveler maximizes his own net benefit, and if the benefit (excluding travel cost) of one traveler does not depend on what others do. A prediction of how the flows will vary if one changes the network (the travel costs), however, does depend on the elasticities or the form of the benefit function; this is what one is usually trying to obtain from such a theory.

Various planning models that are used to evaluate the distribution and the assignment simultaneously rather than sequentially relate the q_{ij} to the costs of travel but they do not explicitly associate an economic benefit with the choice of trips. In such models, the q_{ij} is usually considered to be a function of not only the cost of travel from i to j but also the costs of alternative trip choices.

Even if an economic theory were conceptually correct, one obviously could not, in practice, determine the functions $g_{ij}(q_{ij})$ or their generalization. At best one could only propose some reasonable functional forms for them, involving a few adjustable parameters. The parameters could then be determined so as to give a best fit to some suitable data. This is, in fact, what is done for various distribution or distribution and assignment models.

Here one definitely encounters one of the basic difficulties in transportation planning: The more realistic one tries to make the theory, the more one introduces parameters or functions for which there is insufficient data.

**8.5
The Total Travel
Cost**

In the previous chapters it was assumed that people agree on the costs of trips, and so it seemed evident that they would also agree on the total cost of travel, $fc(f)$ on any link. But if they do not agree on their individual costs of travel, it is unclear whether they would agree on the total cost. Any attempt to arrive at a social-optimal necessarily weighs the benefits of the poor against the rich, and so forth. To have an optimal implies the acceptance of some objective function, a total cost.

In section 8.3, we introduced a hypothetical cost c_3 for not traveling. This is difficult to measure directly and is, perhaps, not even well defined in principle. Actually, one would infer the $f_d(c)$ or the $g(f)$ by supposing that the theory was conceptually correct. Then if one could somehow increase the (agreed upon) cost of travel from a value c to $c + \Delta c$ and observe a decrease in the number of trips that were made, presumably the travelers that chose not to go because of the increased cost must have valued the trip at a cost between c and $c + \Delta c$. According to this interpretation, the travelers with low income but fixed working hours are likely to have the highest cost c_3, but they would certainly find it very difficult to pay the price of losing their jobs.

In formulating an objective function when different travelers did not agree on the cost of choice 3, we did not adopt an average cost and assume that the total cost of choice 3 was the flow f_3 times this average cost. Instead we interpreted the total cost of choice 3 in (8.9) as the sum of each individual's cost *as he saw it*, with equal weight on each increment of flow (trip). It should be noted, however, that the flows f_k were originally defined as quantities that satisfied a conservation principle. We could, in effect, obtain different weights, depending on whether the flows f_k count vehicles or passengers.

In generalizing the results of section 8.3 to situations in which different people have different interpretations of the cost of travel on routes 1 and 2 (route 1 may be private transportation, route 2 public transportation), it is important that we again recognize the shortcomings of postulating that the total cost of travel has the form $f_j c_j(f_j)$, in which $c_j(f_j)$ is a chosen cost (possibly an average over all people or what a planner thinks it should be). A more realistic measure of total cost would result if we simply added the costs for each individual as he sees them.

Conceptually there is not much difference between people disagreeing on the costs of routes 1 or 2 or 3. The cost of choice 3, however, does not depend on how many others choose it (there is no congestion); people disagree only on the value of a single number c_3. In formulating a theory, therefore, it suffices to describe the number of travelers $f_d(c)$ who value the choice more than any number c. Equivalently, one could specify the inverse of this function, the demand $g(f)$. If, however, $c_1(f)$ or $c_2(f)$ vary with f, travelers who disagree would each have their own set of functions $c_j(f)$.

Clearly there is not much point in even trying to formulate a theory in which travelers are selected from a population having arbitrary cost curves because one could not gather the data necessary to use such a theory. To keep the analysis manageable, one must choose a simple population; for example, a one- or two-parameter family of cost curves (along with a description of the probability distributions of parameter values) or just a finite collection of cost curves (along with a description of how many travelers are associated with each). A simple example of a one-parameter family of cost curves $c_2(f)$ would be the family of constants c_2 (analogous to the c_3). One could describe the population of travelers (relative to route 2) by specifying how many or what fraction of them consider the cost of route 2 to be larger than any number c, the analog of the $f_d(c)$, or one could deal with the inverse of this function, the analog of the $c_3(f)$.

As a simple variation on the arguments of section 8.3, suppose that there were a specified number of travelers q that wished to travel from O to D, and they all went (that is, $c_3 > c_1$ or c_2 for everyone). Suppose, also, that travelers agree on the cost of route 1, $c_1(f_1)$, but disagree on the cost of route 2. (Perhaps route 2 is a public transportation system of large capacity that does not become particularly congested.) Everyone agrees that the travel cost on route 2 is relatively insensitive to the flow f_2; they simply disagree on the value of c_2.

By analogy with the scheme of section 8.3, we can order the travelers in increasing order of their values of c_2 and assume that trips are assigned to route 2 in that order. Let $c_2'(f_2)$ denote the cost that the last traveler so assigned associates with route 2 at the flow f_2. This is the analog of the $c_3(f_3)$ in section 8.3, but we will now use a prime as a reminder that this is not a flow-dependent cost function in the usual sense. Only the last traveler assigned to route 2 pays this cost, not all travelers. If we add the costs of travel for each traveler on route 2 as he sees it, the total cost of travel on route 2 would be the analog of (8.9), namely

$$\int_0^{f_2} c_2'(y)dy.$$

If there is an equilibrium between routes 1 and 2, with each traveler taking his cheapest route, the equilibrium

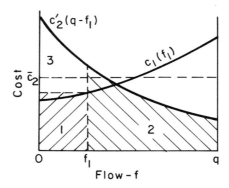

Figure 8.5
A comparison of total costs when
each traveler has his own travel
cost or all travelers have a cost
equal to the average cost.

condition is the usual one,

$$c_1(f_1) = c_2'(f_2), \qquad \text{if } f_1, f_2 > 0,$$ (8.15)

with $f_1 + f_2 = q$. The total cost is

$$T(f_1) = f_1 c_1(f_1) + \int_0^{q-f_1} c_2'(y) dy.$$ (8.16)

Although the last traveler assigned to route 2 sees the costs as being equal on the two routes, everyone else believes that his route is definitely superior to the other.

If we draw a graph of $c_1(f_1)$ and $c_2'(q - f_1)$, as in figure 8.5, we see that, for any value of f_1, the shaded rectangular area 1 represents the cost to trips on route 1, whereas area 2 is the cost for trips on route 2. The total area of regions 1, 2, and 3,

$$\int_0^q c_2'(y) dy,$$

is independent of f_1. By analogy with figure 8.4, we can interpret area 3 as the profit derived by travelers who are allowed to use route 1 instead of 2. To minimize the cost (8.16) or, equivalently, to maximize area 3, one should choose f_1 so that

$$c_1^*(f_1) = c_2'(f_2),$$

and those travelers who consider route 2 more expensive than the marginal cost on route 1 are allowed to use route 1.

The more significant aspects of this model derive from its comparison with assignments and costs for a hypothetical population of identical average travelers as discussed in chapter 6. Presumably, one would interpret the cost of route 2 for an average traveler as

$$\bar{c}_2 = \frac{1}{q} \int_0^q c_2'(y)dy, \tag{8.17}$$

and, if we adopted this as the cost for all travelers, the total cost for an assignment of a flow f_1' to route 1 and $f_2' = q - f_1'$ to route 2 would be

$$T'(f_1') = f_1' c_1(f_1') + (q - f_1')\bar{c}_2. \tag{8.18}$$

If trips were assigned to their cheaper route for this model of identical travelers, the equilibrium would yield

$$
\begin{aligned}
c_1(f_1') &= \bar{c}_2 &&\text{if } c_1(0) < \bar{c}_2 < c_1(q); \\
f_1' &= q &&\text{if } c_1(q) < \bar{c}_2; \\
f_1' &= 0 &&\text{if } \bar{c}_2 < c_1(0).
\end{aligned}
\tag{8.19}
$$

If trips were assigned to minimize (8.18), we would have the corresponding equations, with $c_1(\cdot)$ replaced by $c_1^*(\cdot)$.

Whereas the equilibrium (8.15) corresponds to the intersection of the curves $c_1(f_1)$ and $c_2'(q - f_1)$, as shown in figure 8.5, the equilibrium (8.19) corresponds to the intersection of $c_1(f_1)$ with the constant $\bar{c}_2$. As one can see from this figure, these two equilibrium flows could be significantly different, depending on the shape of the curves $c_1(f_1)$ and $c_2'(f_2)$. In particular, if $c_2'(0) < c_1(q) < \bar{c}_2$, (8.19) would give $f_1' = q$ (no one uses route 2), whereas (8.16) might assign an appreciable fraction of the flow to route 2.

This example illustrates why planners do not use models with identical travelers to predict modal split. Even though the average traveler in most cities obviously must prefer private transportation to public transportation (except under conditions of severe congestion)—the average cost of the former is less than the latter—some people consider public transportation cheaper. It is also true that some travelers prefer urban highways to freeways, scenic routes to commercial routes, and so forth. Any time one tries to predict the partition of travelers between dissimilar routes, one must be conscious of the possibility that some travelers may have special reasons for choosing what may seem to be an inferior route.

That travelers disagree on the cost of travel not only affects what they will actually do but also what a planner would like them to do. The use of (8.18) as an estimate of transportation cost can also lead to incorrect policy recommendations.

In comparing the costs (8.16) and (8.18), we note first that because $c_2'(f)$ is an increasing function of f,

$$\frac{1}{q} \int_0^q c_2'(y)\, dy > \frac{1}{f_2} \int_0^{f_2} c_2'(y)\, dy, \qquad \text{for } 0 < f_2 < q.$$

Therefore

$$f_2\, \bar{c}_2 > \int_0^{f_2} c_2'(y)\, dy, \qquad 0 < f_2 < q,$$

and

$$T'(f_1') > T(f_1), \qquad f_1' = f_1. \tag{8.20}$$

In particular, if f_1' were chosen as the flow that minimizes $T'(f_1')$, $T(f_1')$ would be less than the minimum of T'. The minimum of $T(f)$ would, of course, be generally smaller yet; that is, less than $T(f_1')$.

Basically, the reason for this is that, in (8.16), it is assumed that travelers are assigned to route 2 in order of their preference; thus the average cost to the travelers who use route 2 at any flow f_2 is less than the average cost of route 2 for those who do not use it or for all potential users.

One can also show from figure 8.5 that if each traveler chooses what he considers to be the cheapest route—f_1

satisfies (8.15) and f_1' satisfies (8.19)—then it is still true that $T'(f_1') \geq T(f_1)$.

Even if there were a physical situation in which this theory were conceptually correct—travelers agree on the cost $c_1(f_1)$ but not on the cost c_2—it would be quite difficult to determine the function $c_2'(f_2)$. One could ask each person to state his cost, but most people would not know. Perhaps one could change route 1 so as to shift the equilibrium point and thereby infer the cost to those who switched; but this would presume that everyone also agreed on the new cost of route 1. Thus, even in this simple case, it would be difficult to predict the equilibrium flows or the flows that minimize the total cost. Because travelers are assumed to agree on the cost $c_1(f_1)$, however, they will also agree on the marginal cost $c_1^*(f_1)$. Consequently, they would agree on the tolls that should be charged at any value of f_1 (whatever it may be) to minimize the total cost. Of course, if the travelers disagree on the cost $c_1(f_1)$, the issue of charging tolls becomes much more complex.

From the point of view of design, one of the important consequences of (8.20) and related inequalities is that there is a benefit associated with making routes different in such a way that travelers will disagree on their travel costs. Presumably, if travelers disagree, each will use the route that he likes best and take advantage of the option. An efficient transportation system will, therefore, generally have a variety of facilities (modes), but not very many, because there is usually an economy associated with building facilities so as to attract many trips onto the same route.

Note that, because of these qualitative properties, many of the symmetric networks discussed in chapter 6 and 7 that have many identical routes between the same points are not likely to be very efficient in terms of transportation costs. They exist for reasons relating to land use.

8.6 Generalizations

Attempts to generalize the theory of traffic assignment to allow for differences in traveler preferences have gone in several directions. However, the difficulty in practical modeling is mostly related to the elimination of irrelevant information or parameters that cannot be observed.

A simple generalization of the example in section 8.5 to cases in which travelers disagree on the costs of more than one route would involve quite complex data relating to the sampling of peoples' preferences. Even if the costs of various routes are independent of the flows but each person has his own cost associated with route 1 and route 2, it is not enough simply to know the fraction of people who believe route 1 costs less than c_1 and the fraction who believe route 2 costs less than c_2. One must deal with the joint distributions, the fraction of people who believe that route 1 costs less than c_1 *and* route 2 costs less than c_2. This becomes rather academic because there is no way to collect all the data necessary to infer the values for these functions.

Models usually involving flow-independent costs on various links of a network have been proposed for the joint probabilities of travelers making choices among several routes. Such models are generally considered under the label of "stochastic traffic assignment" models. They involve a variety of assumptions regarding statistical independence of various quantities, special mathematical forms for distributions of costs, and so forth, but contain a few free parameters that can be adjusted to fit any relevant data.

Another category of models (see references 2, 3, 4) are described as "multiclass-user" models. These deal with flow-dependent costs on the links but employ much simpler assumptions regarding the population of users. The users are classified into a few types (classes), with all users of the same class having identical travel costs on all links. The classification would presumably include a separation of cars, buses, and trucks, for example. As a further complication in these theories, however, most models allow the cost to travelers of class j on link k to be a fairly general (increasing) function of the flows on link k for all user classes and possibly even a function of the flows on other links.

The literature on such models deals mostly with existence and uniqueness of equilibrium assignments, the effect of tolls, and possible algorithms for the calculation of flows. Not much attention has been given to the specific form of cost functions that would be used in application and the practical consequences of phenomena associated with multiclass users. It is certainly not yet clear under what conditions it would be necessary to use a multiclass-user model in order to obtain realistic flows, given that considerable data would be needed to use such models.

For multiclass-user models, there is no question that there must exist a system-optimal assignment that will minimize the total cost; that is, the sum of the costs to all travelers as they see it (provided that some assignment is feasible). One would even expect the optimal link flows for each class to be unique for most reasonable link cost functions. One would not expect to achieve this minimum cost, however, by charging tolls, except possibly by charging different tolls to the different classes.

For the user-optimal assignment, the situation is more complicated. For a multiclass-user assignment it is not necessarily true that there exists a generalization of the T^* of chapter 6; that is, some (possibly artificial) global objective that, minimized, will yield a user-optimal assignment for all classes. One can define conditions under which a suitable generalized T^* can be defined, but there is no reason why, for reasonable cost functions, these conditions should be satisfied. It has been shown, at least in hypothetical examples (see reference 2), that for such a general class of models there may be more than one equilibrium assignment, more than one flow pattern having the property that no traveler can find a route that he considers to be cheaper than the one he is on. Whether this is of any practical consequence is, as yet, unclear.

Even for only two user classes, two routes between a single origin and destination, and the travel cost on each route represented as a function of one variable (the total flow or passenger car equivalent), the classification of various types of assignment schemes is quite tedious (see reference 3). There are four cost functions, $c_i^{(j)}(f_i)$, for class j on route i and two input flows, $q^{(j)}$, for users of class j to be split between the two routes. The user-optimal assignment is unique, however, for given cost functions and $q^{(j)}$ in the cases for which the $c_i^{(j)}(f_i)$ are functions of only one flow variable.

In such models it is possible, particularly for sufficiently small values of the $q^{(j)}$, that all class 1 users will definitely prefer route 1 and class 2 users will prefer route 2 (or vice versa, or they could both prefer route 1 or both prefer route 2). Each will, of course, use what he prefers under a user-optimal assignment. If the flow of class 1 increases, the cost on route 1 will increase until class 1 users find the costs on the two routes to be equal. As the flow increases still further, the class 1 users will divide between the two routes so as to maintain the equality of their travel costs.

This will uniquely determine the total flow on each of the two routes, $f_1 + f_2 = q^{(1)} + q^{(2)}$ and $c_1^{(1)}(f_1) = c_2^{(1)}(f_2)$, and, therefore, the costs $c_1^{(2)}(f_1)$, $c_2^{(2)}(f_2)$ for users of class 2. These two costs will, generally, not be equal; therefore, class 2 users will all be on one route (route 2 in this case).

It is typical of the multiclass-user-optimal assignments that a flow from one class of users split between two routes will dictate the costs for other users. All others will then be on one route or the other. Even in the system-optimal assignment, it will usually be true that all of one class are assigned to one route, although, for certain ranges of the flows $q^{(j)}$, it may be that flows for more than one class are split between the same routes.

It is not clear yet whether it is easier or more instructive to deal with discrete classes, as in the multiclass-user assignment, or a continuum of classes, as in the stochastic traffic assignment (but with flow-dependent costs). The former would appear to be more manageable, but the discreteness also introduces some artifical complications resulting from the fact that an equilibrium of costs between two routes involves all (identical) users of the same class, whereas for a continuum of classes nearly all users definitely prefer one route. It may be computationally easier to shift flows iteratively to preferred routes than to hunt for the user class that is dictating the route costs. Certainly this whole area of problems involving nonidentical travelers is still in a rather primitive state of development.

Problems   1
Suppose there is a known demand function $g(f)$ describing the travel demand between two points, O and D; that is, $g(f)$ is the cost for which a flow f would find it profitable to travel. There are two routes from O to D. One route has a cost per trip $c_1(f_1)$; the other has a cost per trip $c_2(f_2)$ for all trips using the second route if the flow is f_2. For purposes of illustration, let

$$g(f) = 6 - f, \qquad 0 \le f \le 6,$$
$$c_1(f_1) = 1 + f_1,$$
$$c_2(f_2) = 2 + f_2.$$

(Cost could be in units of dollars, flows in units of trips per second.)

a. If an operator of both facilities could charge a single (nonnegative) toll for travel between O and D (the users would be free to choose the cheaper route), what toll would he charge so as to maximize

i.   the benefit to society; that is, the sum of the benefit to travelers and the toll collected;

ii. the toll collected;

iii. the benefit to the travelers?

b. If the operator could charge separate (nonnegative) tolls on the two routes, what tolls would be charged to maximize (i), (ii), or (iii)?

2
Show that $T'(f_1') \ge T(f_1)$ if f_1 satisfies (8.15) and f_1' satisfies (8.19).

References   1
Daganzo, C. F., and Sheffi, Y. "On Stochastic Models of Traffic Assignment," *Transportation Science* 11 (1977): 253–274.

2
Netter, M. "Equilibrium and Marginal Cost Pricing on a Road Network with Several Traffic Flow Types," *Proceedings of the Fifth International Symposium on the Theory of Transportation and Traffic Flow*, pp. 155–164. American Elsevier, 1972.

3
Jeevanantham, I. "A New Look at the Traffic Assignment Problem," *Proceedings of the Fifth International Symposium on the Theory of Transportation and Traffic Flow*, pp. 131–154. American Elsevier, 1972.

4
Dafermos, S. C. "The Traffic Assignment Problem for Multiclass-User Transportation Networks," *Transportation Science* 6 (1972): 73–87.

9 OPTIMAL NETWORK GEOMETRY, NONCONVEX OBJECTIVE FUNCTIONS

The final phase of the transportation planning procedure is supposed to give a comparison of costs, benefits, and so forth, of various proposed transportation networks designed to serve a projected trip generation, presumably as a basis for recommendations as to what facilities should be built. Transportation planners are, however, reluctant to propose any specific objective that would formally make this an "optimization problem," and for good reasons.

Because it is possible to estimate costs of construction, maintenance, and operation of various facilities and costs of travel on any network for a given set of O-D flows, it should, in principle, be possible also to formulate an optimal (cheapest) schedule of construction to accommodate any projected evolution of the O-D flows. But (unfortunate from the point of view of any mathematical theory) the construction of new facilities affects the pattern of land use, thus the trip generation, and also the trip distribution (O-D flows). Ultimately the decision as to what should be built involves questions about the social life pattern, economy, and politics of the population. In the context of this optimization problem, one would ask: What evolution of the O-D flows would the people want if, through policy decisions regarding transportation or zoning regulations, the O-D flows could be changed? There is, of course, no simple answer to this question. The goals of a community, as reflected by the majority view, tend to change over time as political and social issues change. Many a proposed transportation plan, supposedly based on the best available predictions of travel patterns and costs, has been rejected by voters (often through rejection of a bond issue), which implies, of course, that some planner misjudged the community's objectives.

The whole transportation planning procedure is built, more or less, around a traditional theme that the city planners will determine the projected growth of a city, the land use, and other social patterns to make an estimate of the O-D flows, and the transportation planners (or engineers) will then design an efficient transportation system to accommodate this traffic. Actually, the procedure is not quite this simple because the distribution models (but not the generation models) involve the transportation costs. Thus an "evaluation" of a proposed network includes an estimated redistribution of trips but not usually a quantitative measure of the benefits associated with the redistribution (which is inherently not very well defined). In any case, the procedure usually bypasses a rather basic question as to whether one should provide (cheap) transportation to allow easy access from any origin to many destinations or design land use so as to minimize the need for (expensive) transportation.

To describe any future network, one must specify an enormous number of parameters, such as capacities, locations, and types of all facilities that one might build. Ideally, the planning procedures should evaluate the advantages of all possible proposals, but, because the analysis of even a single proposal (including distribution, assignment, and so forth) requires very lengthy calculations, the final phase usually compares only a few selected choices of parameter values.

That one performs a complete network evaluation only for a few proposals implies that someone must have made some preliminary calculations to determine the many parameters associated with each proposed network. This must involve a suboptimization of some sort, a selection of the most appropriate parameter values within some class of proposals in order to single out one for detailed evaluation. Again the logic of this is not completely rigorous because a proposed suboptimal design, evaluated on the basis of some assumed traffic flow pattern, is used as a basis for a complete network evaluation from which a new traffic pattern will emerge (with new O-D flows). The proposed design may then no longer be a suboptimal design for the new flow pattern. One could try to iterate the calculations—evaluate a new proposed network based on the revised O-D flows—but there is no guarantee that the iterations would converge. Indeed there is no guarantee that if one builds certain facilities, travelers will use them in the manner for which they were planned. One can decide what to build, but one has only limited control over how it is used (what O-D flows will evolve).

Although there are some rather ill-posed questions associated with the general problem of how one should select a network to accommodate the multifaceted objectives of society, some research has been done on the reasonably well-defined problem: Given a projected evolution of the O-D flows, the costs of travel, construction, operation, and so forth, determine a schedule of construction and mode of operation so as to minimize the total transportation-related costs. These costs would, presumably, be the sum of discounted costs; that is, any cost incurred at time t weighted by some factor of the form e^{-rt} for some constant r (the discount rate).

This problem, which may, in principle, be well posed, is quite difficult to solve. It has not yet received a great deal of attention as compared with the more popular problems discussed in previous chapters, probably because no one has found any very effective ways of solving the general problem or even special cases of it. (Most of the literature on optimal network design deals with computer algorithms for solving various special types of problems. A fairly extensive review of this is contained in reference 1.)

Not very much progress has been made on the general time-dependent problem of determining both when and what to build. In an abstract sense, this is a dynamic programming-type problem: Given the present state of the system, determine what action should be taken next to minimize the (discounted) costs over the (infinite) future, assuming that one will also use an optimal strategy at all future times. Because there is an economy of scale in construction, one builds facilities in discrete units rather than continuously in time, but the set of possible strategies is generally of astronomical dimensions and the costs of various strategies are extremely complex functions of the parameters.

Most of the literature on strategies for network expansion deal with the static optimal problem: Given the O-D flows, assumed to be stationary over a sufficiently long period of time, and the costs of construction of a certain specified set of possible links, determine the links that minimize the cost (per unit time) of transportation plus construction; or, given a budget for construction, determine the links that minimize transportation costs for the given budget.

We will be concerned here mostly with the mathematical nature of this problem and some of its qualitative properties rather than with algorithms for computation. If or when some efficient numerical schemes are obtained, they will need to exploit special features of this particular problem; the set of all strategies is too large for a computer to handle by "general purpose" search techniques. We will also be limited mostly to the analysis of certain static optimal strategies, not because they are necessarily good approximations to dynamic optimal strategies but because a theory of dynamic optimal strategies has not been developed to the point that one can say very much. There is no question, however, that a dynamic optimal pattern of construction will generally be such that the optimal network existing at any time is never the same as the static optimal pattern associated with the O-D flows for that time.

Because it is expensive to discard or move facilities and there are economics of scale in construction, it is typically advantageous to overbuild certain facilities in anticipation of future needs and future construction. If one must build a new bridge, for example, one may build it with a capacity even exceeding that of the approaches, knowing that it cannot be used efficiently at first. When the demand becomes sufficiently large, one will enlarge the approaches but will not need to build another bridge. At any time, a dynamic optimal network will generally contain certain congested facilities that are to be expanded in the near future and other recently built facilities that are underutilized. Such "unbalanced" systems would not appear in a static optimal geometry.

Also, if one chooses a certain spacing between arterials, freeways, or transit lines that may be optimal at one level of demand, there is not much one can do if the demand increases and one wants a different spacing. Routes once built are seldom moved.

In the previous discussions, particularly in chapter 6, it was assumed (or implied) that the cost of travel on any route or link was the cost of making a trip compared with not making a trip on a facility that would not be changed in response to an increase in travel. Thus the cost as a function of the flow was considered to be increasing (even convex) as a result of congestion. If, however, one includes the cost of construction, operation, and so forth, and further asks what sort of network one should have, one is, in effect, introducing a more general cost function for travel that admits the facility might change as more trips are added. The facility will not necessarily become more congested as the flow increases; on the contrary, the (expanded) facility may become more efficient with a lower average cost per trip.

We shall see that there are certain abstract similarities between the traffic assignment problem discussed in chapter 6 and the problem of optimal construction, operation, and assignment. One of the main reasons why even the static optimal network design problem is mathematically and computationally more difficult than that discussed in chapter 6 is that the objective function in this more general problem is not necessarily a convex function of the assignment. The objective function may have many local minima, which means that a computer algorithm (other than some general search procedure) will not necessarily converge to the true minimum.

**9.2
Total Cost of
Transportation**

We have already assumed in chapter 6 that the cost of travel for a single trip on a route of given design can be represented as the sum of the travel costs on all links of the route and the cost of travel on any link is a function of the flow on only that link. Neither of these assumptions is quite correct (particularly the latter), but we will continue to use them. Although we did not explicitly specify the numerical values of these costs, we did postulate some qualitative properties of them, and it was fairly clear what some typical choices would be.

If we wish to consider construction, maintenance, and operating costs, we could generalize this by first introducing some abstract (multidimensional) space of facility types. The space is to be defined so that, by specifying a point in this space, one will have described any and all parameters relevant to a description of costs. As in chapter 6, we still wish to separate from the total system cost the travel costs of individual trips because, for any specified network, we must still make some assumptions as to which routes travelers will take. Therefore, the cost of travel on link (i, j) will now be considered a function $c_{ij}(f_{ij}; z_{ij})$, in which z_{ij} represents some point in this space of facility types.

Here we are already making another simplifying assumption that the cost of travel on link (i, j) depends on the properties of the link (i, j) and the flow on (i, j) but not on the properties of other links, z_{kl}, such that $(k, l) \neq (i, j)$. This seems to be a reasonable assumption. It does not mean, however, that there are no dependencies between the links. There may be some subsidiary constraints on the set of z_{ij} that require adjacent links to have similar features (for example, on a bus route, the headways between buses must be the same on all links of the route), just as there are constraints on the f_{ij} (for example, they must satisfy the conservation equations).

For any given choice of the z_{ij}, the $c_{ij} (f_{ij}; z_{ij})$ are assumed to have properties relative to the f_{ij} described in chapter 6. The total travel cost will also have the same form as the T of chapter 6:

$$T = \sum_{(i, j) \in L} f_{ij} c_{ij}(f_{ij}; z_{ij}). \tag{9.1}$$

These costs are considered to be costs per unit of time (day or year). If we wish to add the construction and operating costs, they must also be converted into the same units. Although costs of construction or of a bus fleet may actually be lumped costs incurred at the time of investment, we shall suppose that one can borrow money at a rate of interest, thus these costs will be interpreted as the (daily or annual) interest cost.

We will not be concerned here with the actual numerical values of these costs or how they might be decomposed into various components. The total cost of owning, operating, and using a facility for the link (i, j) will simply be decomposed into two parts. One part is the cost (per unit time) that one incurs whether one uses the facility or not; this is a cost for $f_{ij} = 0$ that we will usually call "construction cost" (or possibly "operating cost" if we are considering a bus route) regardless of what it actually includes. The remaining flow-dependent part of the cost (which now, by definition, vanishes for $f_{ij} = 0$) will be assumed to have the form $f_{ij}c_{ij}(f_{ij}; z_{ij})$.

There may be certain maintenance and/or operating costs that are increasing functions of the flow f_{ij}, parts of which must therefore be absorbed into the flow-dependent part of the total cost. What we, in effect, are assuming is that the cost $c_{ij}(f_{ij}; z_{ij})$, on the basis of which travelers select their routes, includes (possibly through gasoline tax or fares) all flow-dependent components of the total cost. We could obviously consider a more general type of cost structure with two flow-dependent parts, but we are more concerned here with the illustration of certain issues than with quantitative accuracy. We do not want to introduce any more functions than necessary.

The total cost associated with a link (i, j) will thus be considered to have the form

$$C_{ij}(z_{ij}) + f_{ij}c_{ij}(f_{ij}; z_{ij}), \tag{9.2}$$

in which by definition, $C_{ij}(z_{ij})$ is the "construction cost," the cost for $f_{ij} = 0$. The total transportation system cost of the network is assumed to be additive on the links and, consequently, has the form

$$T_s = \sum_{i, j} [C_{ij}(z_{ij}) + f_{ij}(p_{ij}; z_{ij})]. \tag{9.3}$$

This implies that the construction cost of the link (i, j) depends only on its type z_{ij} but not on the z_{kl}, $(k, l) \neq (i, j)$, an assumption that is not necessarily correct if there is an economy associated with building two or more links simultaneously. Actually, there is usually an economy associated with building a two-way link compared with two one-way links. It may, therefore, be advantageous to consider a slightly more general cost function than (9.3), a single cost for the two links, (i, j) and (j, i), rather than separate costs for each.

The space of facility types is assumed to contain the "null facility," which we can label as $z_{ij} = 0$; it costs nothing, $C_{ij}(0) = 0$, and carries no flow, $c_{ij}(f_{ij}; 0) = \infty$. We may imagine, therefore, a graph of nodes and links that includes anything we might conceivably build. Any actual graph (subgraph) can then be identified by the set of $z_{ij} \neq 0$.

If a facility (z_{ij}) already exists and one is concerned with what new facilities are needed to accommodate some future trip distribution, then the cost of the existing z_{ij}, $C_{ij}(z_{ij})$ is irrelevant. Whether one is paying interest on past investments, one cannot presumably change the existing $C_{ij}(z_{ij})$, even if one abandoned the link, $z_{ij} = 0$ (at no cost). For such facilities, one could interpret the $C_{ij}(z_{ij})$ in (9.3) as the cost of replacing the existing facility by one of any new type z_{ij} (thus 0 if there is no change or the facility is abandoned).

The (static) "optimal network design problem" can be interpreted as the following: For any given (stationary) O-D flows q_{ij} between nodes i and j and traffic assignment scheme such as described in chapter 6, determine the set of z_{ij} (subject to possible constraints) that minimizes (9.3). One can, of course, define appropriate generalizations of this with less restrictive assumptions about the form of the cost functions, including multiclass users. The immediate issue is whether one can obtain solutions of this problem and, if so, what properties the solutions have.

We can make one important preliminary observation. A real transportation network contains a very large number of different origins and destinations; for example, individual residences. If construction costs nothing and one wished to build a network to minimize (9.3), one obviously would build a facility $(z \neq 0)$ on essentially every possible link. In the extreme situation in which one can build anywhere, the network that minimizes transportation costs alone would undoubtedly have a separate high-speed route between every O-D pair. A city would be solid pavement with multilevel crossings, and each trip would travel a straight-line (minimum distance) route from its origin to destination. The most important qualitative aspect of the cost (9.3) is that there is, generally, an economy of scale. One does not build a separate route for each trip. Because there are many origins and destinations, each having access to the network, the optimal network will, on one hand, contain many feeder links (driveways, residential streets, sidewalks) that carry low flows and generally have low speeds. On the other hand, one wishes to focus many routes (with different origins and destinations) through the same links to take advantage of economies in construction or operation. The economy is further exaggerated by the fact that facilities of large capacity (freeways, transit lines) also have high speed; that is, small $c(f; z)$.

The network of optimal design basically represents a balance between an increase in travel distance (from straight line paths) against a greater economy (perhaps even shorter trip time) from routing trips onto the same links. Unfortunately, this fact is often buried in mathematical details or even lost in attempts to convert this optimization problem into one that is easier to solve.

**9.3
User versus System
Optimal
Assignment**

In chapter 6 we considered two assignment schemes, one in which each user chooses a route of minimum cost to himself and the other in which trips are assigned so as to minimize the total transportation cost T. For each of these assignment schemes there is an associated optimal network design problem. If, for each possible network (set of z_{lm}), trips are assigned to routes according to some scheme, one, in effect, determines the f_{ij} in (9.3) as functions of (all) the z_{lm}. The T_s should then be minimized with respect to the z_{lm}, with each f_{ij} changing as the z_{lm} change. Because the evaluation of the f_{ij} is a rather complex problem for any reasonable size network with specified z_{lm} (as described in chapter 6), the evaluation of the T_s as a function of the z_{lm} and the calculation of the minimum with respect to the z_{lm} appears to be horrendous.

For a user-optimal assignment (which seems to be the more realistic scheme), a solution of this problem does, indeed, appear to be rather hopeless. For any given network, the user-optimal assignment can be determined by minimizing the T^* with respect to the f_{ij}, subject to the usual conservation equation constraints, or by minimizing an artificial objective function with respect to the f_{ij} (for each fixed z_{ij}), subject to the same constraints:

$$T_s^* = \sum_{i,j} \left[C_{ij}(z_{ij}) + \int_0^{f_{ij}} c_{ij}(y; z_{ij}) dy \right]. \tag{9.4}$$

This is like a noncooperative game between users and planners. For any fixed z_{ij}, the users try to minimize T_s^*, but the planners counter by choosing the z_{lm} so as to minimize T_s. In the second assignment scheme, however, both the f_{ij} and the z_{lm} are chosen so as to minimize the same objective, T_s. Because of this, the latter problem is much easier to solve and is perhaps even manageable, as we shall see (see also references 2 and 3). Another potentially manageable problem would be to minimize T_s^* with respect to both the assignment and the z_{lm}. Although this may have no economic significance, a solution of this artificial problem may be useful for comparisons.

The optimal network for a user-optimal assignment is, of course, not necessarily the same as for a system-optimal assignment. If one could determine and build the latter network but then allow travelers to choose the cheapest routes, some travelers might find that the route of least cost is not the same as the route of least marginal cost. One could evaluate the user-optimal assignment for this system-optimal network by the methods discussed in chapter 6 and evaluate the extra cost incurred from allowing travelers to choose the cheapest routes. If this extra cost were significant, an optimal network for the user-optimal assignment would tend to modify the network so as to reduce the opportunities for travelers to select routes of high marginal cost (as illustrated in chapter 7).

As a practical matter, it seems rather unlikely that this would be a very serious issue. Because an optimal network will tend to focus trips onto certain links in order to justify the construction of economical and high-speed facilities, the routes of least marginal cost are likely to be unique (for most trips) and also of least cost.

An evaluation of flows for either the user- or system-optimal assignment is complicated because the f_{ij} must satisfy very awkward constraints induced from the (multicommodity) conservation equations, not because the functions T or T^* are particularly difficult to evaluate. There may also be some constraints on the z_{lm}. If, for example, the network is a public transportation system and the z_{lm} describe flows of buses on the links, the z_{lm} would also satisfy conservation equations analogous to the f_{ij}. If, however, the network is for private transportation, the parameters z_{lm} could include the settings of traffic signals. A change in the green phase of a N-S signal, for example, would induce a change for the E-W signal. In principle, this optimization of the z_{lm} includes the problem of optimal settings of signals on a network, but the travel costs introduced in planning models usually do not contain this much detail. It is presupposed that the signals (if any) on main roads are synchronized so that one can define a typical average speed; signals on minor roads (if considered) are not synchronized and have a lower average speed.

The trick that one must exploit in order to reduce the system-optimal network design problem to potentially manageable size is to reverse the order of minimization with respect to the f_{ij} and z_{lm}. Whereas we have been describing this problem as if one would first determine the f_{ij} for a fixed network using the methods of chapter 6 and then minimize the T_s with respect to the z_{lm} (as one must do for a user-optimal assignment), if both the f_{ij} and z_{lm} are chosen so as to minimize the same function T_s, one could first minimize T_s with respect to the z_{lm} for fixed f_{ij} and then minimize T_s with respect to the f_{ij}. It is easier to perform the minimization first with respect to the z_{lm} because the constraints (if any) on the z_{lm} are generally simpler than those on the f_{ij}. (All these minimizations must be done subject to any applicable constraints on the f_{ij} and z_{lm}.)

9.4 Unconstrained Facility Choice

It seems unrealistic to assume that for a two-way link the cost $C_{ij}(z_{ij})$ is unrelated to the cost $C_{ji}(z_{ji})$. For purposes of illustration, we will assume here that all links are two-way, that a link is labeled either (i, j) or (j, i) but not both, and that $C_{ij}(z_{ij})$ represents the combined cost of (i, j) and (j, i).

To avoid unnecessary complications with the assignment, the f_{ij} will be interpreted as 24-hour flows. It is then reasonable to assume that as many trips go one way as the other, $q_{ij} = q_{ji}$ and $f_{ij} = f_{ji}$. We can then combine the equal terms $f_{ij}c_{ij}(f_{ij}; z_{ij})$ and $f_{ji}c_{ji}(f_{ji}; z_{ji})$ as twice one or the other, provided that the cost functions are equal in the two directions. Actually, this is a rather crude hypothesis because it implies that the assignment depends on the symmetric 24-hour flows rather than the morning or evening peak flows, which are likely to be asymmetric.

Suppose that there are no constraints on the z_{lm}. One can build anything on the (symmetric) link (l, m), independent of what one builds on any other two-way link. For any fixed values of the f_{ij} compatable with the conservation equations, a particular z_{lm} appears only in the (l, m) term of (9.3). To minimize T_s with respect to z_{lm}, it suffices, therefore, to minimize this one term. The (l, m) term of T_s also involves only the flow f_{lm} on that link; thus for a given flow f_{lm} on link (l, m), one should build the most efficient facility to serve this flow (independent of what one does elsewhere). If we let

$$C_{lm}^*(f_{lm}) = \min_{z_{lm}}[C_{lm}(z_{lm}) + 2f_{lm}c_{lm}(f_{lm}; z_{lm})] \qquad (9.5)$$

for each two-way link (l, m), then the minimization of T_s reduces to determining an assignment with flows f_{ij} that minimizes

$$T_s^\circ \equiv \min_{z_{lm}} \sum_{i, j} [C_{ij}(z_{ij}) + 2f_{ij}c_{ij}(f_{ij}; z_{ij})]$$

$$= \sum_{i, j} C_{ij}^*(f_{ij}) \qquad (9.6)$$

(in which the sum is over the two-way links), subject to the usual conservation constraints on the f_{ij}. If one can determine the $f_{ij} = f_{ij}^\circ$ that minimizes (9.6), the optimal network can be easily determined as the z_{lm}° that defines the minimum in (9.5) at $f_{lm} = f_{lm}^\circ$ for each l, m.

Formally, the minimization of T_s° appears to be the same type of problem encountered in the traffic assignment: Minimize a sum of functions of the flows f_{ij}, subject to the (multicommodity) conservation equations. The only apparent difference is that a function $C_{ij}^*(f_{ij})$ replaces the functions $2f_{ij}c_{ij}(f_{ij})$ or $2\int^{f_{ij}}c_{ij}(y)dy$ in the T or T^*. Although no specific mathematical form was assumed for the $c_{ij}(f_{ij})$ in chapter 6, there is one critical difference: The terms of T and T^* could reasonably be assumed to be convex, but $C_{ij}^*(f_{ij})$ certainly is not. For certain idealized situations, one could even consider the $C_{ij}^*(f_{ij})$ to be concave.

One can devise computer algorithms similar to those described in chapter 6 that are guaranteed, at each stage, to decrease the value of T_s° and, consequently, to converge at least to a local minimum. For example, one could define a "generalized marginal cost" of assigning a trip along some route R (plus another trip in the reverse direction) that is additive on the links of R:

$$\sum_{(l,\,m)\in R} dC_{lm}^*(f_{lm})/df_{lm}, \tag{9.7}$$

This new marginal cost includes what was previously the marginal cost plus the incremental cost of building an optimal facility to accommodate one unit of flow. The terms of (9.7) can also be used to define a cost (distance) on links so that the shortest path algorithms can be employed to determine the route of minimum cost in (9.7) from any node i to node j.

Any reassignment of trips to cheapest routes in terms of (9.7) is certain, initially, to cause a decrease in T_s°; there are various methods of iteration that cause T_s° to decrease at each stage. There is no guarantee, however, that this "steepest descent" method will lead to the global minimum of T_s°. If one can make an intelligent guess at a starting assignment, perhaps it will.

The nature of the assignment that minimizes T_s° is likely to be quite different from that described in chapter 6 for T or T^*. Stationary points for T_s obtained by setting the derivatives of T_s with respect to route flow assignments equal to zero (by equating marginal costs of travel on used routes between the same origin and destination), are likely to give local maxima for T_s rather than local minima. Optimal assignments are now most likely to lie on a constraint boundary in the space of the f_{ij}. In particular, because we started with a network that potentially included a link anywhere that one might conceivably build a facility, the optimal network will be one for which f_{ij} and $C_{ij}^*(f_{ij})$ are equal to zero for most (z_{ij}); one does not build anything in most of the places where one could have built something.

If a network includes all links on which one might build something, it will obviously have a very large number of links (in any real applications). The computation problem is not simple; one cannot usually guess at a good trial solution. Even a simplified version of the problem that specifies what facility to build on a link if anything is built yields an "integer programming" problem; for each link one must determine whether to build the facility. Such problems are, unfortunately, very difficult to solve, in general. To make any progress one must exploit various qualitative features of the cost functions; that is, one must understand some of the issues.

**9.5
Cost on a Link**

Before one can even start the minimizations of T_s° in (9.6), one must evaluate the $C_{lm}^*(f_{lm})$ from (9.5), which involves a minimization of the link cost with respect to all facility choices z_{lm} for all values of f_{lm}. Despite the fact that the space of all facility types may be quite complex, involving many parameters and including facilities ranging from a driveway to a freeway or a bus route to a transit line, this minimization is, as a practical matter, not a difficult part of the problem (at least if there are no constraints on the z_{lm}).

For any particular link (l, m), one could draw curves of

$$C_{lm}(z_{lm}) + 2fc_{lm}(f; z_{lm}) \tag{9.8}$$

as a function of f for many typical facility types, such as two-lane roads, three-lane roads, and freeways. For any single facility z_{lm}, this function has a positive value $C_{lm}(z_{lm})$ at $f = 0$ (the "construction cost"). Near $f = 0$, it has a slope $c(0; z_{lm})$, the trip cost (per trip) at zero flow. For higher values of f, the slope is the marginal cost $c^*(f; z_{lm})$, which typically increases with f and becomes infinite as f approaches the capacity of the facility z_{lm}.

For each f, the $C_{lm}^*(f)$ is the minimum of all such curves;

that is, the lower envelope of all curves (9.8). Although there may be a very large number of facility types, one should have little difficulty in guessing which parts of which curves (9.8) are worth drawing in order to construct at least an approximate curve $C_{lm}^*(f)$. For various types of highways, the facility that is cheapest to build (unpaved road, or two-lane road)—the one with the lowest C_{lm}—usually has the steepest slope (slowest speed) and least capacity. The cost curves for the larger facilities cross those of the smaller facilities and become the cheaper facility for flows above some intersection point, as illustrated schematically in figure 9.1. Some actual numerical estimates are given in references 2 and 4, but these curves will, of course, vary from year to year (due to inflation or improvements in highway technology) and from city to city.

The flows in question here could range from about 1 car per day to the order of 10^5 per day. To show the numerical values of $C_{lm}^*(f)$ over this range, it would perhaps be best to draw the graphs on a log-log scale, but in discussing the mathematical properties of $C_{lm}^*(f)$ we prefer to consider the curves as if they were drawn on the usual linear scale.

Although the individual curves of (9.8) for fixed z_{lm} are generally convex, the envelope $C_{lm}^*(f)$ tends to bend in the opposite direction. Because the transition from a two-lane road to a three-lane road, for example, is not continuous, this envelope may show a scalloped shape, as in figure 9.1; it may be convex over a certain range of flows but then have a discontinuity in slope when the facility type changes. At sufficiently high flow values corresponding to multilane freeways, the curve $C_{lm}^*(f)$ may cease to bend downward because one cannot continue indefinitely to gain efficiency by adding more and more lanes.

For most ranges of flow, if seems reasonable to assume that, for any z, one could build a facility z' consisting of two identical neighboring facilities of type z. Furthermore, the facility z' would cost twice as much as z and have the same cost per trip at twice the flow:

$$C_{lm}(z') = 2C_{lm}(z) \quad \text{and} \quad c_{lm}(2f; z') = c_{lm}(f; z).$$

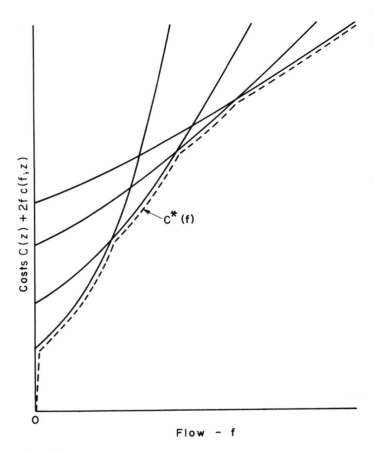

Figure 9.1
Cost curves for various types of
facilities (solid lines) and the
minimum cost (broken line).

Therefore

$$C_{lm}(z') + 2(2f)c_{lm}(2f; z') = 2[C_{lm}(z) + 2fc_{lm}(f; z)]. \quad (9.9)$$

If, for some flow f_0, the facility z were the optimal facility,
then the right-hand side of (9.9) would be $2C^*_{lm}(f_0)$. The
minimum cost $C^*_{lm}(2f_0)$ to accommodate a flow $2f_0$ can be
no larger than the cost for the facility z' to carry $2f_0$;
therefore,

$$C_{lm}^*(2f_0) \le 2C_{lm}^*(f_0).$$

More generally, for $n = 2, 3, \ldots,$

$$C_{lm}^*(nf_0) \le nC_{lm}^*(f_0). \tag{9.10}$$

The condition (9.10) does not imply concavity of $C_m^*(f)$, although concavity would imply (9.10), but it does imply that the curve $C_{lm}^*(f)$ becomes, in some sense, less steep as f increases.

These arguments apply only if the facilities z and z' are new facilities. If a facility exists initially on the link (l, m), one could still construct the curve $C_{lm}^*(f)$ for all new facilities as described. One could also draw the curve (9.8) for the existing facility z_0 in which the $C_{lm}(z_0)$ is set equal to zero or, more correctly, simply equal to the cost of maintenance for the facility at zero flow (but not the investment cost). The correct $C_{lm}^*(f)$ would now be the minimum of these two curves. This new cost curve would be likely to follow the cost curve for the existing facility for low flow values (because of the low value of $C_{lm}(z_0)$). This portion of the cost curve is, of course, convex. At some flow, however, the curve for the new facilities will cross that of the existing facility. For flows above the crossing it pays to abandon the existing facility and build a new one.

The possibility that one might retain the existing facility but build a new facility parallel to the existing facility can be handled most easily (at least formally) by adding a separate link to the network from l to m on which one can build anything. The traffic assignment scheme will then automatically (in principle) partition the flow between the two links and indicate whether one will actually build something $(z \ne 0)$.

It should be noted here that, as a practical matter, the details of this minimization with respect to the z_{lm} cannot be taken too seriously. Perhaps it is more appropriate to interpret the "optimal facility" for flow f as that which is customary or socially acceptable. There are many intangibles that enter into the choice of facility type. In residential areas, for example, one might find (according to direct cost estimates) that it is cheapest to build a one-lane road to accommodate very low flows; yet, for various reasons, such as parking, esthetics, or convenience, this is seldom done. It is clear also that freeways are frequently built in places where they appear to be unnecessary from an economic point of view (but necessary from a political point of view).

Although it is implied in this formulation that one must determine the optimal facility separately for each link (l, m), in practice one would also classify the costs $C_{lm}^*(f)$ as a function of certain physical characteristics of the link (l, m). Clearly the cost $C_{lm}^*(f)$ depends on the physical length of the link, the cost of the right-of-way (particularly whether the link is in an urban, residential, or rural area), and the type of junctions.

Because any link can be artificially subdivided into various sections and because we have assumed here that the cost of any link is independent of what one builds elsewhere, it has, in effect, already been implied that the cost of a homogeneous link is an additive function on its subsections and thus a linear function of its length. If we neglect the cost of junctions or assume that the number of junctions is (more or less) proportional to the length, then we can also assume that the cost of a link is proportional to its length.

For a link (l, m) of length L_{lm}, we can write

$$C_{lm}^*(f) = L_{lm} C^*(f), \tag{9.11}$$

in which $C^*(f)$ is the minimum cost for a facility of unit length to carry a flow f. This cost $C^*(f)$ will still depend somewhat on the location of the link (l, m), particularly whether it is in an urban or rural area, but we shall not include this dependence explicitly in the notation. The important point here is that one need only evaluate the $C^*(f)$ for several types of physical regions in order to determine the $C_{lm}^*(f)$ for all possible links.

9.6 Public Transportation Systems

The problem of determining an optimal design for a public transportation system has certain abstract similarities to that of a highway system, but there are also important differences. It has already been noted that for a public transportation system the parameters z_{lm} would include the headways between vehicles. Because the vehicles usually traverse more than one link, there will be conservation constraints associated with the z_{lm}. There are other complications; for example, the speed of travel depends on the frequency of stops. Although we will not formulate a realistic scheme for determining an optimal geometry of a public transportation system, we will illustrate that the issues of concavity of the objective function with respect to the assignment of trips exists here also (even more so).

If the public transportation system is a bus system using the existing road network, construction cost (in the usual sense) is not significant. The cost analogous to the $C_{lm}(z_{lm})$ in (9.2) is essentially the operating cost. It is reasonable to assume that there is a certain cost per unit distance c_0 for operating a bus (if bus stops are equally spaced). For a link of length L_{lm}, the operating cost per unit time, therefore, should be approximately

$$c_0 L_{lm}/h_{lm},$$

in which h_{lm} is the (time) headway between buses.

The other flow-dependent part of the travel cost is the components, one of which represents the cost equivalent of the passengers' riding time. This part should be (nearly) independent of the headways or even the flow f_{lm} of passengers on the link (l, m). The total riding-time cost, therefore, could be represented by a term of the form $f_{lm}c_{lm}$, where c_{lm} is the riding cost per trip.

The other flow-dependent part of the travel cost is the cost equivalent of the time that passengers spend waiting for a bus. This must be paid on entering the system and at each transfer point. It is this part of the cost that makes the routing of trips by way of public transportation and the design of the network quite different from that of a highway network.

If each trip on the link (l, m) must wait an average of half a headway to use this link, and the cost per unit time of waiting is c_w per passenger, then the total waiting cost per unit time on this link would be

$\frac{1}{2} c_w h_{lm} f_{lm}.$

The combined cost of operation, waiting, and riding per unit time, therefore, has the form

$$\frac{c_0 L_{lm}}{h_{lm}} + \frac{1}{2} c_w h_{lm} f_{lm} + f_{lm} c_{lm}, \qquad (9.12)$$

which is still consistent with the general form (9.2). The total cost on the network would be the sum of this over all links.

If the link (l, m) were a shuttle bus route running only between l and m, then one could disregard the possible conservation equations associated with buses on adjacent links. If all vehicles have sufficient capacity to accommodate anyone who wishes to board, one can also operate with (essentially) any headway. The headway h_{lm} now assumes a role analogous to the facility type z_{lm} of section 9.5.

For given flow f_{lm} we can minimize (9.12) with respect to the h_{lm} by setting the derivative of (9.12) with respect to the h_{lm} equal to zero. The optimal h_{lm} is determined from the equation

$$\frac{c_0 L_{lm}}{h_{lm}^2} = \frac{1}{2} c_w f_{lm},$$

and the corresponding minimum cost on the link (9.12) is

$$C_{lm}^*(f_{lm}) = [2 c_0 c_w L_{lm} f_{lm}]^{1/2} + f_{lm} c_{lm}. \qquad (9.13)$$

If the headways are equal in both directions on the link (l, m), the flow f_{lm} should be interpreted here as the sum of the flows in the two directions.

Although the details of this formulation are very artificial, the feature of (9.13) that should be emphasized is the dependence on the flow f_{lm}. The first term of (9.13) is proportional to $f_{lm}^{1/2}$, which makes $C_{lm}^*(f_{lm})$ a concave function of the flow.

If the complete objective function T_s for a public transportation system is a concave function of the link flows (which is typically the case) and there is more than one possible route for the trips between any origin and any destination, then the

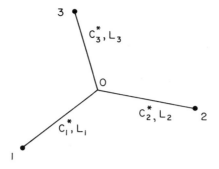

Figure 9.2
Network junction.

optimal assignment of these trips will definitely be such that they are all assigned to the same route. One will never obtain an optimal assignment by equating trip costs on two or more routes, which was typical of the assignment schemes of chapter 6 for a convex objective function.

**9.7
Network
Junctions**

Although the optimal network will involve a compromise between longer paths and higher speed made possible by shared facilities, this fact is obscured in the formulation of the previous sections; it would eventually emerge from the calculations through the selection of some relatively small subset of the original network on which one builds facilities with nonzero capacities. The collection of all possible facilities, however, is extremely large; in fact, there is a multidimensional continuum of facility locations. One can use analytic procedures at least to determine some local geometries.

The final network will certainly involve some junctions. Suppose, for example, that a network was constructed to join points 1, 2, and 3 of figure 9.2 at a junction at 0. Let the costs per unit length of travel plus construction on the three legs be C_1^*, C_2^*, and C_3^*. These are considered to be given and independent of the direction of the links from j to 0. If the lengths of the links from j to 0 are denoted by L_j and we disregard the cost of the intersection itself, the cost of the networks is

$$T_s = C_1^* L_1 + C_2^* L_2 + C_3^* L_3. \tag{9.14}$$

The point 0 should be located and the values of the L_j determined so as to minimize T_s. The L_j are, of course, subject to the constraints that the network pass through the points 1, 2, and 3.

The mathematical form of (9.14) is the same as the potential energy of a system of three strings tied at 0, with tensions C_1^*, C_2^*, and C_3^*. The equilibrium configuration of a string system would also be one of minimum potential energy. There is no point in setting derivatives equal to zero to locate the point 0; we would only be redeveloping the mathematics that were invented to treat conveniently the corresponding problem in mechanics. If we let $\overline{C_j^*}$ be a vector of magnitude C_j^* in the direction from 0 to j, then the condition for a minimum of T_s is that the vector "forces" $\overline{C_j^*}$ at 0 add to the null vector:

$$\overline{C_1^*} + \overline{C_2^*} + \overline{C_3^*} = \overline{0}. \tag{9.15}$$

If the magnitudes C_j^* are known and each C_j^* is larger than the sum of the other two, a triangle of sides C_1^*, C_2^*, C_3^* satisfying (9.15) will determine the angles between the $\overline{C_j^*}$, which are, in turn, the angles between the vector lines from j to 0.

Knowing the angles between the L_j, we must still locate the point 0 and determine the values of the L_j. One fast way to do this is to draw three lines, making the appropriate angles, on a sheet of tracing paper. Slide this paper over a paper showing the three points 1, 2, 3 until one can make the three lines pass through the three points (if possible). (One can also locate the point 0 analytically, but the formulas are somewhat cumbersome.) In surveying, this is known as the "three point problem": From a point 0, a surveyor measures the angles between three known benchmarks at 1, 2, and 3 and then calculates his position.

It may happen that the C_j^* do not form a triangle (because one C_j^* is larger than the sum of the other two) or, if they do form a triangle, one cannot join the points 1, 2, 3 to 0 with these angles (because one of the angles of the triangle 1, 2, 3 is too large). But, of course, the minimization of (9.14) may be such that the point 0 is at one of the points 1, 2, or 3 and, consequently, cannot be found by the equilibrium of forces (by setting derivatives equal to zero).

A simpler way to find the location of the point 0, including the special cases, is to locate the points 1, 2, and 3 on a board and drill holes through the board at these points. Attach weights C_1^*, C_2^*, and C_3^* to strings, pass the strings through the holes, and tie the ends together. When the knot is released, it will move to the optimal location. If the weights pull the knot to one of the holes, that is the optimal location.

If four or more links should meet at 0, the vector forces must still balance at 0, but the problem of locating 0 so that the links pass through given points 1, 2, 3, and 4 becomes analytically very tedious. A string model, however, will give a solution very quickly. By measuring the lengths of the strings, one can also calculate T_s so as to compare the cost with other possible geometries.

If the links are highways, there may be some objection to having routes meet at odd angles. But if one were to introduce additional penalties for badly angled intersections and costs for having a curved road with various radii of curvature, one would still find that long highways (joining cities, for example) would have the directions described, except near the intersection, where the highways would bend to meet at right angles or merge angles.

For local streets the situation may be quite different. One would not necessarily curve a driveway just to obtain a direction that minimized length. If intersections on city streets occur frequently, the costs incurred from the intersections may be more important than the costs proportional to length of road. Because one four-way intersection is likely to cost less than two T-intersections, one usually builds four-way right-angled intersections and a compatible street system of parallel streets.

Locating the point 0 so as to minimize the cost T_s was a convex problem. For given values of the C_j^*, T_s is a convex function of the L_j. The concave problem described in the last sections related to the flow dependence of the C_j^*. One

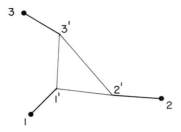

Figure 9.3
Optimal geometry of a network
joining three points.

can see the interplay between these concave and convex
aspects of the optimal network by considering a more
general problem: Given three points 1, 2, and 3, an O-D
table with flows $q_{ij}(i, j = 1, 2, 3)$, and a cost function $C^*(f)$
(the same everywhere), determine the network that will
accommodate the flows q_{ij} and minimize the total cost. The
geometry need not be the same as in figure 9.2.

Imagine building a network with an arbitrary number of
links connecting the points 1, 2, and 3 to each other and/or
to any number of other nodes one may wish to add. If,
however, the function $C^*(f)$ is concave, we know that the
flow q_{ij} between i and j will all be assigned to the same route
in the optimal geometry. For simplicity, let us assume that all
links are two way and that C^* is actually a function of the
total two-way flow (equivalently $q_{ij} = q_{ji}$) so that, in effect,
there are only three routes used 1-2, 1-3, and 2-3.

One can easily prove that the optimal network joining
three points 1, 2, 3 must have a shape such as that shown in
figure 9.3 (or some limiting case thereof) in which one or
more of the new nodes 1', 2', 3' coincide with 1, 2, or 3, or
with each other. The argument follows. There are only two
routes leaving point 1, one goes to 2, the other to 3. A first
point 1' is located where these two routes either separate or
meet the route from 2 to 3. Wherever the point 1' may be,
the link (1, 1') must be a straight line because f and $C^*(f)$
are specified on the link (1, 1') and the straight line
minimizes the length (and thus the cost) of (1, 1'). If, at 1',
the routes from 1 meet the route from 2 to 3, then all three

routes have a common point. The minimum-cost routes from a common junction would be of the type shown in figure 9.2 because one can do no better than to send all trips from the common junction to node j along the same route. If, on the other hand, point $1'$ is where the routes from 1 separate (without joining the route from 2 to 3), then the route from $1'$ to 2 must follow a straight line until it meets the route from 2 to 3 at $2'$.

Actually the triangle $1'$, $2'$, $3'$ of figure 9.3 must, in all cases, either collapse to a single point 0 or be so large that one of its points coincides with 1, 2, or 3 because all costs are proportional to lengths and independent of direction or position. For any triangle $1'$, $2'$, $3'$, one can construct another network with the same angles between links but a larger (or smaller) triangle $1'$, $2'$, $3'$. The total cost would vary linearly with the size of $1'$, $2'$, $3'$ and would have a minimum either for the smallest or largest triangle that one could build. Whatever the final configuration may be, the angles between links at any junction point must satisfy an equilibrium of vector "forces" such as (9.15). We will not try to enumerate all the possible shapes of optimal networks and the conditions under which they occur; we are more interested in some of the qualitative consequences of the concave cost functions.

Although we have not evaluated the solution of this three node problem explicitly, it has been reduced to a comparison of, at most, a few possible candidates. The key step in the analysis was the recognition that, for a concave cost function, all trips from i to j would use the same route. Suppose, however, we attack the problem in the manner of section 9.3. If we introduce some finite but very large number of subsidiary nodes (on a lattice, for example), connect them together in all possible ways, and then assign traffic to various routes on the network so as to minimize T_s, we, of course, find that the flows on nearly all links are zero; that is, the links are not built. As with most concave problems, however, we might find that a slight change in the q_{ij} would cause the assignment suddenly to jump from one route to another, not necessarily even a neighboring route.

Sometimes it is easier to solve a problem in a space with

an infinite number of points than with a finite number of points. This analytic method of solution actually allowed for a junction at any point 0, 1', 2', 3' in the two-dimensional continuum, not just at a finite set of possible points. Instead of comparing assignments to a discrete set of routes, we compared assignments to routes infinitesimally close together. Thus in locating the point 0 of figure 9.2, we, in effect, compared an assignment of the flow to routes passing through 0 with assignments to a route through a neighboring value at 0', for example. If we had considered a network that was the superposition of all networks, as in figure 9.2, with all possible locations of 0, the assignment of the flow to minimize T_s would automatically have put all the flow on the routes passing through the optimal 0 and nothing on the other routes.

If one had programmed a computer to search for an optimal assignment on a network with a finite but large number of nodes and gave no special instructions as to where to look, the computer might not even come close to the optimal network in a reasonable computation time. It would go through some very lengthy procedure of putting links in the network, taking them out, etc., trying to find the best combination in an astronomical number of possibilities.

Unfortunately, one cannot solve problems with four or five nodes analytically (in a reasonable time) nor can the computer solve them very well by a search procedure. Anything we can learn from the analysis of simple problems might at least give some qualitative properties of desirable (if not optimal) networks.

Some interesting effects can be seen even if we consider the simple special case of the three node problem of figure 9.3 in which the points 1, 2, 3 form an equilateral triangle and $q_{12} = q_{13} = q_{23} = q$ (everything is symmetric with respect to permutations of 1, 2, and 3). In this special case, there are two obvious candidates for the optimal network, namely those shown in figures 9.4a and 9.4b.

Because all costs are proportional to length, the scale of the triangle 1, 2, 3, is irrelevant; it suffices to make the sides of the triangle in 9.4b equal to 1/3. The cost of construction plus travel for 9.4a and 9.4b are $3^{-1/2}C^*(2q)$ and $C^*(q)$, re-

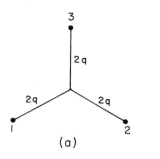
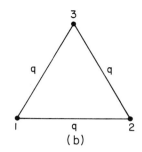

Figure 9.4
Two possible networks joining
three points.

spectively. The total length of the network (a) is $3^{-1/2} \approx 0.58$, compared with 1 for (b), but the length of travel per trip on (a) is $2/\sqrt{3} \approx 1.15$, compared with 1 for (b).

From a graph of $C^*(q)$ one can easily identify those values of q for which (a) is cheaper than (b), or vice versa. The function $3^{-1/2}C^*(2q)$ is essentially the same graph as $C^*(q)$, except for a change in the vertical and horizontal scales.

If there is a nonzero cost $C^*(0)$ for $q \to 0$, then (a) is cheaper for sufficiently low q because it is the network of minimum length. Suppose, however, that $C^*(q)$ increases linearly with q over a sufficiently large range of q. For example, $C^*(q)$ might consist of the fixed construction cost $C^*(0)$ plus a constant cost c per trip mile of travel so that $C^*(q) = C^*(0) + qc$. The cost of (a) is then

$$3^{-1/2}C^*(2q) = 3^{-1/2}C^*(0) + 3^{-1/2}2\,qc,$$

and the cost of (b) is

$$C^*(q) = C^*(0) + qc.$$

Although the graph of $3^{-1/2}C^*(2q)$ has the smaller intercept at $q = 0$, it has the steeper slope. When the total cost is dominated by travel cost, (b) is preferred because it has the shortest routes.

As q increases, however, one may find it advantageous to change facility types; the cost $C^*(q)$ may experience a sudden decrease in slope. It may then be desirable to build a

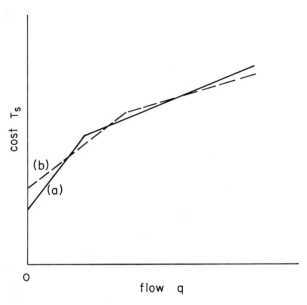

Figure 9.5
Comparison of costs for the
networks of figure 9.4.

network of shorter length but higher speed; network (a) may
be preferred again. There is nothing to prevent the curves
$3^{-1/2}C^*(2q)$ and $C^*(q)$ from crossing many times. This is
illustrated schematically in figure 9.5. In the present example,
the ratio of slopes $2/\sqrt{3} \approx 1.15$ is not very large. In figure
9.5 this difference in slope is exaggerated.

One can easily show that the optimal networks for this
symmetric three node problem actually are of the type shown
in figure 9.4a or 9.4b and that the optimal geometry does
jump discontinuously from one to the other as q changes;
there are no optimal geometries between (a) and (b) through
which one can make a continuous transition. This type of
behavior is typical of many nonconvex optimization
problems. If we had considered the assignment of flow to a
network consisting of the superposition of (a) and (b), we
would find that all q_{ij} is assigned to the same route, but for
certain values of q_{ij} the assignment suddenly jumps from one
route to another.

This discontinuous behavior of the static optimal network geometry occurs very often and is quite important in planning for a growing traffic demand. If one builds the network in 9.4a for a present value of q and later finds that the network in 9.4b is better, there is no way of obtaining (b) except by abandoning what has already been built. The dynamic optimal strategy is likely to require that one builds two legs of the network (b) initially and the third later, even though the two leg network is not optimal at any time.

9.8
Major-Minor Road
Junctions

The following problem is closely related to that of the previous section. Suppose that one has a main road carrying through-traffic f in each of the east and west directions. From a point O off this main road, one wishes to construct a feeder road(s) that will carry a flow f_1 to and from points to the west of 0 and a flow f_2 to and from points to the east of 0. Unlike the previous example, one does not wish to change the direction of the main road at the junctions with the feeders, perhaps, in part, because f_1 and f_2 are small compared with f.

One has the option of building two feeder roads at angles θ_1 and θ_2, as in figure 9.6a, one for f_1 and the other for f_2, or a single feeder for the flow $f_1 + f_2$ at the angle θ, as in figure 9.6b. The determination of the optimal angles θ_1, θ_2, or θ is similar to that described in section 4.3. If the cost per unit length of route is $C^*(f)$ at flow f, then the incremental cost to the main road traffic of unit displacement of the junction at θ_2 is

$$-C^*(f_2 + f) + C^*(f).$$

The extra cost to the feeder route is $C^*(f_2) \cos \theta_2$. The optimal θ_2 is therefore

$$\cos \theta_2 = [C^*(f_2 + f) - C^*(f)]/C^*(f_2). \tag{9.16}$$

We, of course, expect the right-hand-side of (9.16) to be less than 1. If it is considerably less than 1, then θ_2 is close to $\pi/2$.

In 9.6b, the cost on the main road of a unit displacement of the junction is

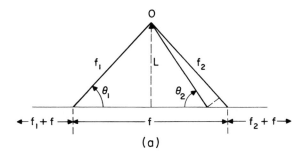

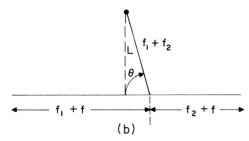

Figure 9.6
Feeder roads to a major road.

$$-C^*(f_2 + f) + C^*(f_1 + f).$$

The optimal angle θ is given by

$$\cos \theta = [C^*(f_2 + f) - C^*(f_1 + f)]/C^*(f_1 + f_2). \qquad (9.17)$$

If $f_1 = f_2$, then $\theta = \pi/2$.

The extra cost of having the origin of trips at O (a distance L from the main road) rather than on the main road ($L = 0$) is

$$\frac{LC^*(f_1)}{\sin \theta_1} - L[C^*(f_1 + f) - C^*(f)]\frac{\cos \theta_1}{\sin \theta_1} + \frac{LC^*(f_2)}{\sin \theta_2}$$
$$- L[C^*(f_2 + f) - C^*(f)]\frac{\cos \theta_2}{\sin \theta_2}$$

for the geometry of 9.6a. At the optimal values of θ_1, θ_2, this cost becomes

$$LC^*(f_1) \sin \theta_1 + LC^*(f_2) \sin \theta_2.$$

The corresponding cost for the geometry in 9.6b is

$$LC^*(f_1 + f_2) \sin \theta;$$

consequently, the geometry in 9.6b is preferred if

$$C^*(f_1 + f_2) \sin \theta < C^*(f_1) \sin \theta_1 + C^*(f_2) \sin \theta_2. \tag{9.18}$$

In particular, if $f_1 = f_2$, this simplifies to

$$C^*(2f_1) < 2C^*(f_1) \sin \theta_1$$

or, by virtue of (9.16),

$$\left[\frac{C^*(2f_1)}{2C^*(f_1)}\right]^2 + \left[\frac{C^*(f_1 + f) - C^*(f)}{C^*(f_1)}\right]^2 < 1. \tag{9.19}$$

The conditions (9.18) or (9.19) will usually be satisfied; it is generally cheaper to build one road to carry a flow $f_1 + f_2$ than to build two roads to carry flows f_1 and f_2. Certainly, in (9.19), we expect to have $C^*(2f_1) < 2C^*(f_1)$ as in (9.10). The second term of (9.19) is expected to be appreciably less than 1 because it should be cheaper to carry a flow f_1 on the main road than on a separate road; that is, $C^*(f_1 + f) - C^*(f) < C^*(f_1)$.

9.9 Distributed Origins Unlike the examples in the previous sections, an important feature of real transportation systems is that the origins and destinations are usually distributed over a wide area so finely that one would like to think of them as having a density (on a scale of distance comparable with the average trip length). As a consequence of this, any optimal road network must have a hierarchy of facility types, such as driveways, city streets, and arterials. There must be a driveway (or sidewalk) for each individual origin or destination, a city street to collect traffic from driveways, arterials to collect traffic from city streets, and so forth.

Unfortunately, even in hypothetical systems with highly symmetrical O-D patterns, optimal networks may be very complicated; despite any symmetry of the O-D table, there may be no identical streets. If, for example, one were to postulate an O-D table with rotational symmetry about a city center, or translational symmetry (over a large but finite region), the optimal network will not show the same symmetry. Obviously, the only strictly rotationally symmetric road network would have solid pavements; if there were a street at one angle θ, there would be streets at every θ. Again, it is the concave behavior of $C^*(f)$ that rules this out; one does not build a network with an infinite number of routes, each carrying an infinitesimal flow.

It is not even clear that the optimal network will have any symmetries. The circularly symmetric O-D table may lead to an optimal network that is not even invariant to rotations through discrete angles of $2\pi/n$ for an integer n. A translationally symmetric O-D table may not give an optimal network that forms any lattice structure invariant to translations through specified distances. The principle that almost rules out any symmetries for a concave objective function is that between any two points that are possible origins and destinations of trips there must be a unique route to which these trips will be assigned so as to minimize the total cost. Thus if one postulates a rectangular grid of roads, any trip that could travel (from one corner to the opposite diagonal corner) by way of more than one route would find one route better than others; the flows (and the costs) would be unequal on parallel roads.

No one has made much progress toward finding optimal networks. Efforts in this direction have been confined mainly to a comparison of various (nonoptimal) alternative patterns. This is an important problem, however, because there are many types of networks (telephone lines, power lines, sewer lines, street systems) that occur in our environment for transporting various things, and the same problem arises in all of them. If one cannot solve the problem, one should at least try to eliminate systems that are clearly inferior to other systems.

The optimal network for transporting something from distributed sources to a single destination, or vice versa, will obviously look (in three dimensions) qualitatively like the root system of a tree (perhaps nature has found its own solution for gathering water and food from a distributed source). This is obviously rather complicated. Perhaps it is easier to start with a system having a higher degree of symmetry in two dimensions and an O-D table with translational symmetry in two directions plus symmetry with respect to 90° rotations and reflections; that is, a distribution of both origins and destinations.

Suppose we start from the driveways to the city street. According to the analysis of the last sections, there will be a three- or four-way intersection between a driveway and the street. If the driveway trips are equally likely to turn right or left, the optimal angle of the intersection will be nearly a right angle (perhaps exactly a right angle, four-way intersection if two driveways meet on opposite sides of the street) because the flow on the driveway is small compared with the city street. Strict application of the results of section 9.7 might require that the city street zigzag slightly, diverting slightly for each driveway. This, obviously will not do (at this level of flow) because of esthetics or because the cost of bending a street is not negligible (that is, the costs are not exactly proportional to length, independent of shape).

How one builds driveways does not seem to be of great importance in network design, but certain qualitative features of this are important because the same features repeat for the other hierarchies. Driveways are cheap per mile but expensive per trip-mile. If one were to build a network to carry at least a certain minimum number of trip-miles of travel, one would not build only driveways (a single route between every O-D pair), despite the low cost per mile. Driveways are inefficient but necessary because each origin and destination must have at least one link.

Driveways are usually built by property owners, not the government, and individuals, for various personal reasons, may build all sorts of odd things (but neither are they inhibited by social prejudice from choosing some logical schemes). It is not clear exactly how we should define a driveway. Formally, one should perhaps define it as the first link one uses in leaving an origin, but there are many physical things that would qualify under this definition. The usual notion of a driveway is a link that connects a single origin with a link that is shared by others (a city street). Because construction cost is generally large compared with user cost for this link, a person who owns land adjoining two streets will not ordinarily build two driveways. There is only one direction of travel from the origin. Even though the trip may have a destination opposite to the direction of the driveway, one is willing to pay the extra penalty of traveling in the wrong direction. Clearly the overriding consideration is to get onto a shared facility as quickly as possible. This means: Go to the nearest one and meet it perpendicular.

Some people park their cars on the city street. One could consider the sidewalk as the first link. Again, it may be opposite to the direction of the trip. (One could also argue that there is a sidewalk to the driveway and this is likely to be in a direction unrelated to the trip direction. One should also note that the sidewalk is likely to be perpendicular to the driveway, which, in turn, is perpendicular to the city street).

There are other facilities between driveways and city streets, however, so it is not clear where to separate the two; a driveway can be shared by several origins. There is not much geometrical difference between a shared driveway and a dead-end street, or between an alley and a through street. By sharing a driveway (alley, or whatever) the construction cost is distributed among several users. The one shared facility is cheaper than several separate facilities. One could worry about whether a common driveway should zigzag to meet its users in an optimal way (it often does) or whether the driveway should vary in width or quality from one end to the other. Sometimes the quality does vary; a paved road degenerates into a dirt road, but this is not always economical. Construction cost is not strictly pro-portional to length and the sum of costs of its parts. There is an economy in uniformity. It may be just as cheap to extend a paved road to the user at the end of the street as to build a dirt road near the end.

The choice between dead-end streets or through-streets is usually based on esthetics, land use, etc., rather than just transportation. From the point of view of transportation alone, the through-street has the advantage of giving users an option of starting journeys in either direction. This is to be balanced against construction (including land) cost. In many cases the end of a dead-end street will not be far from the end of another dead-end street. It would not cost very much to convert two dead-end streets into one through-street. The choice is between the cost of the link and the direction flexibility. The more users that share the street, the more one should favor the through-street. Custom seems to favor the through-street, but this may be a legacy from the time when walking was the usual mode of transportation.

The issues in building driveways relate mainly to schemes to gain access to more efficient facilities. The choice may disregard the destination of the trips completely or make minor compromises by providing two directions of travel (usually not more than two). It was implied throughout the discussion, however, that the lengths of the trips were large compared with the lengths of the driveways (alleys, side-walks).

The complete network will contain a hierarchy of sub-networks, each new level being coarser, faster, and of larger capacity than the previous one. There may even be several levels in this hierarchy before one reaches a scale of spacing between links comparable with the mean trip length. Actually, the distribution of trip lengths is rather broad; links at the lower levels serve a dual purpose of providing the complete route for short trips but access to the next level for long trips. If, however, all trips were of comparable length, the issues at the lower levels would be quite different than at the higher levels. At the lower level, the network is designed to give economical access to the next level (independent of the destination of trips). At the top level, the issue is to provide efficient paths between points near the origin to points near the destination (from which the trip is completed at a lower level).

Until one reaches a level for which the link spacing is comparable with the trip length, the issues in connecting one level with the next are nearly the same at all levels. If all trips were eventually to use freeways, the design of city streets should follow the same type of logic used in the design of driveways. At each level the origins would appear as if they were located on a finer grid from the level below.

The city, county, or whatever, would be seen on a coarser and coarser scale as one moves up the hierarchy.

Reference 5 of chapter 7 gives a fairly detailed analysis of optimal spacings for a multiple-hierarchy, rectangular grid highway system serving a trip distribution with translational symmetry in two directions but fairly general trip length distribution (illustrated, however, mostly with an exponential distribution of trip length). The function $C^*(f)$ is approximated by a piecewise linear function generated by cost curves typical of residential streets, arterials, and freeways.

Although the mathematical details of this are rather tedious, the procedure is conceptually straightforward. As in previous illustrations, the difficult problems generated by the concave objective function are bypassed if one admits only a finite parameter family of possible networks. For the example in reference 5, one admits rectangular networks created by the superposition of identical parallel roads of arbitrary spacing at each hierarchy. This does not generally yield the true optimal geometry even for a translationally symmetric O-D table but it does have right-angle junctions and should give a network with a cost quite close to the minimum.

Because there are only a few unknown parameters (the transverse spacings between roads), the optimal values of these parameters can be evaluated by elementary calculus. Typical of concave problems, however, the optimal geometry changes discontinuously as the trip intensity varies. In particular, at some traffic level, one may suddenly find it advantageous to upgrade all the roads in the hierarchy and simultaneously increase the spacing or suddenly introduce a new hierarchy.

As illustrated in section 7.8, the optimal spacings of any hierarchy is basically a balance between the construction cost of that hierarchy and the travel (access) cost on the next lower hierarchy in the orthogonal direction, but it is nearly independent of the travel cost on the former or the construction cost of the latter.

For a public transportation system, the nature of an efficient transportation system is quite different. For a highway system the penalty for a right-angle turn on a route is so small that it is usually neglected in the overall cost, but the cost of a transfer between routes of a public transportation system is a very prominent issue. A trip on a public transportation system should not involve more than one or two transfers. Although there may be a hierarchical structure of sorts, the network will have only a few hierarchies (buses and transit lines). An efficient network is likely to have a geometry with trips focused on a few transfer points rather than a grid structure with transfers at each intersection of the grid. Some comparisons of different geometries are given in reference 5.

Problem   A bus company operates express bus service to the city from depots located at each end of a street of length D. There is a uniform density of residences along this street and they generate ρ trips per unit length per unit time. The trip time from the depot to the city is the same from each depot. The bus company maintains a fleet of vehicles, each of capacity C at both depots, but will dispatch a vehicle only when it is full.

Passengers must walk to the depot at a speed v and wait for the next dispatch; average waiting time is equal to half a headway. Each passenger will choose to go to the depot that minimizes his average trip time.

For many years, the bus company provided equal service from each depot following this strategy, but the demand ρ gradually decreased each year. Then one year all the passengers shifted to the depot at O and the bus company had to discontinue service from D. What happened?

Because this resulted in a further loss of customers, a traffic consultant was hired to give advice. The consultant predicted that if the company reduced the capacity of the vehicles or dispatched them when three-quarters full, service could be restored at both depots. Would you accept this advice?

References   1
Steenbrink, P. A. *Optimization of Transport Networks.* New York: John Wiley & Sons, Inc., 1974.

2
Steenbrink, P. A. "Transport Network Optimization in the Dutch Integral Transportation Study," *Transportation Research* 8 (1974): 11–27.

3
Newell, G. F. "Optimal Network Geometry," *Proceedings of the Sixth International Symposium on Transportation and Traffic Theory* (Sydney, Australia), pp. 561–580. Sydney: A. H. and G. W. Reed, 1974.

4
Tanner, J. C. "The Comparative Evaluation of Idealized Road Networks" *Beiträge zur Theorie des Verkehrsflusses, Proceedings of the Fourth International Symposium on the Theory Traffic Flow and Transportation,* pp. 166–175. Karlsruke, 1968.

5
Newell, G. F. "Some Issues Relating to the Optimal Design of Bus Routes." *Transportation Science* 13 (1979): 20–35.

INDEX

Page numbers in italics indicate figures.

Date Due
